Oil and the Outer Coastal Shelf:
The Georges Bank Case

Oil and the Outer Coastal Shelf

The Georges Bank Case

William R. Ahern, Jr.

The Rand Corporation

Ballinger Publishing Company ● Cambridge, Mass.
A Subsidiary of J.B. Lippincott Company

Library of Congress Catalog Card Number: 73-12300

International Standard Book Number: 0-88410-303-X

Printed in the United States of America

Library of Congress Cataloging in Publication Data

Ahern, William R
        Oil and the outer coastal shelf.
        Originally presented as the author's thesis, Harvard.
        Bibliography: p.
        1. Petroleum in submerged lands–Georges Bank. 2. Environmental policy–Georges Bank. I. Title.
TN871.3.A33 1973   553'.28'0916346   73-12300
ISBN 0-88410-303-X

Table of Contents

List of Figures   vii

List of Tables   ix

Foreword   xi

Preface   xiii

Chapter One
The Decision Problem and Possible Outcomes   1

Chapter Two
Oil and Gas under Georges Bank   7

Physiography and Geography   7
Industry Activities   8
Estimates of Recoverable Petroleum Under Georges Bank   9
Expected Possible Production from Georges Bank   12
Conclusions   14

Chapter Three
Economic Aspects of Possible Outcomes   17

Future Sources of Oil and Gas   17
Effects of Georges Bank Production on Imports   21
Prices and Costs of Imported and Georges Bank Petroleum   23
Balance of Payments Considerations   28
Other Domestic Energy Sources   34
Federal Revenue   34
Impacts on New England Regional Economy   35
Transportation Costs   38

Chapter Four
Environmental Aspects of Potential Oil and
Gas Development on Georges Bank   45

Introduction   45
Predicting Numbers and Sizes of Possible Oil Spills   48
The Life of Spilled Oil   55
Effects of Oil on Marine Life   63
Effects of Oil in Water on Human Health   74
Possible Surface Movement of Oil Slicks Originating from
      Georges Bank Petroleum Development   77
Environmental Impacts of Transportation of Georges Bank   94
      Oil and Gas to Shore
Coping with Oil Spills   101

Chapter Five
Major Findings and Recommendations Based on
Economic and Environmental Considerations   109

Summary of Findings   109
Recommendations   113

Chapter Six
Legal and Political Considerations   115

Legal Considerations   115
Political Considerations   117

Sources Consulted   125

Index   131

About the Author   135

List of Figures

1-1   Georges Bank Orientation Map   3

1-2   General Descriptive Model Linking Decision on Leasing Georges Bank with Possible Outcomes   5

2-1   Area Potentially Favorable for Oil and Gas Deposits Under Georges Bank   8

3-1   Past and Projected Energy Sources   18

3-2   U.S. Gas Supply-Demand Balance, 1966-1990   20

3-3   Impact of Production of Medium and High Estimates of Potential Georges Bank Oil and Gas, On Stream in 1980   22

3-4   Long-Run Impact on U.S. Balance of Payments from Investing in a 100,000 Barrel a Day Production Capacity in the Middle East and from Importing 100,000 Barrels a Day for the First Year and for Twenty-Two Years after the Investment   32–33

4-1   General Descriptive Model Linking Georges Bank Oil and Gas Development with Impacts on the Environment   46

4-2   Sizes of Thirteen Large OCS Oil Spills   52

4-3   Poisson Probability Distributions for Numbers of Oil Spills on Georges Bank over a Forty Year Production Period   53

4-4   Numbers of OCS Oil Spills, 1964-1971, by Month   55

4-5   Volume of Spill, Average Film Thickness after Period of Initial Spread, and Area Covered   57

4-6   Factors Affecting Marine Organisms   62

4-7	General Divisions and Characteristics of the Marine Environment	64
4-8	Hypothetical Slick Movement	77
4-9	General Circulation Directions	78
4-10	General Late Spring and Early Summer Water Movements off the Northeast Coast of the United States	79
4-11	General Water Movements off the Coast of the Northeast United States During Late Summer, Autumn, Winter, and Early Spring	80
4-12	Monthly Wind Rose Charts-Georges Bank	81
4-13	Launch Points for Oil Spill Path Simulation in MIT Study	83
4-14	Total Catch, ICNAF Subarea 5, 1954-1970	87
4-15	Total Haddock Catch, ICNAF Subarea 5, 1952-1970	88
4-16	Habitats of Major Species on Georges Bank	90
4-17	Possible Oil Pipeline and Tanker Destinations	96
4-18	1970 Oil Tanker and Barge Traffic at New England Ports	97

List of Tables

2-1   McKelvey Estimates of Potential OCS Oil and Gas Resources   9

2-2   Nelson and Burke Estimates of Potential OCS Petroleum
        Resources   10

2-3   Five Estimates of Atlantic OCS Potential Recoverable
        Petroleum   10

2-4   Estimates of Potential Recoverable Petroleum under Georges
        Bank   11

2-5   Estimates of Potential Average Annual Petroleum Production
        from Georges Bank   11

2-6   Reduced Set of Estimates of Potential Average Annual
        Petroleum Production   13

2-7   Estimates of Potential Average Daily Petroleum Production
        from Georges Bank   13

3-1   Components of the Price of a Barrel of Crude Oil Landed at the
        U.S. East Coast   24

3-2   Comparison of "Who Gets What" from the Buying of a Barrel
        of Georges Bank Production or of Saudi Arabian Crude   26

3-3   Comparison of "Who Gets What" in Importing from Saudi
        Arabia or in Producing from Georges Bank Thirty-Five Million
        Barrels of Oil a Year   27

3-4   Dollar Flows for Hypothetical Incremental Imports of One
        Million Barrels per Day   29

3-5   Dollar Flows for Hypothetical Incremental Imports of 100,000
        Barrels of Oil from the Middle East   30

3-6   Potential Federal Revenues from Georges Bank Petroleum Development and from Importing an Equivalent Amount of Oil   36

3-7   Typical Schedule to Produce Proved Reserves of One Billion Barrels in the North Sea   40

3-8   Summary of Possible Economic Impacts on New England Associated with Potential Petroleum Production from Georges Bank   41

4-1   List of Major OCS Oil Spills, 1964-1971   47

4-2   Major OCS Oil Spills by Year, Number of Fixed Producing Structures, Total Number of Wells, and Annual Production   49

4-3   Major Contributing Causes of the Sixteen Major OCS Oil Spills   50

4-4   Expected Number of Spills on Georges Bank, Forty Year Period   53

4-5   Possible Numbers of Oil Spills from Georges Bank Oil Production Over a Forty Year Period   54

4-6   Characteristics of Fractions of Oil   56

4-7   Numbers of Major Oil Spills That Would Each Have a Five Percent Chance of Reaching Shore Over a Forty Year Period   84

4-8   Probabilities of Numbers of Spills Reaching Shore from Georges Bank Oil Operations During a Forty Year Period   85

4-9   Nominal Catch in ICNAF Division 5Ze by Species, Total and U.S. Catch   88

4-10 Total Catch by Country in 1970, ICNAF Division 5Ze   89

4-11 Net Change in Risk to New England of Nearshore Oil Spills Resulting from Transport of Georges Bank Oil and Gas   100

5-1   Summary of Possible Outcomes   110

Foreword

William Ahern's analysis of potential petroleum development off the East Coast of the United States is one of the first doctoral dissertations to emerge from the new Public Policy Program in Harvard's Kennedy School. The Program offers analytic tools in four fields: decision theory, economics, statistics, and politics, along with a fifth field devoted to applications, and a sixth concerned with substance. Those who seek the Ph.D., as Mr. Ahern did, are examined in these fields and then do a dissertation, usually a problem-solving exercise, applying tools to substance. This study is an early, excellent example. It is distinguished by three qualities that make it, in my view, a model piece of work. It comes to grips with a timely and important problem in the real world. It is comprehensive, clear, imaginative and persuasive. And it was completed under pressure of a short deadline. It should be helpful to all those interested in off-shore petroleum development, and to all concerned with the art of public policy analysis.

Richard E. Neustadt
Professor of Government
and Associate Dean

Kennedy School of Government
Harvard University
July 1973

Preface

This book is my response to an assignment from the Public Policy Program at Harvard's Kennedy School of Government. I had one year to complete the analysis of a messy, complex, real-life government policy problem.

My topic was risky: policy problems can easily blowup on you. There might not be time to get all the relevant information in one year, a year that had to include writing, typing and proofing, the final report. Happily, the topic area was narrowed by the offer of financial support from the Program on Marine Policy and Ocean Management at Woods Hole Oceanographic Institution on Cape Cod. The issue had to deal with the sea.

It was clear from regular reading of the *Boston Globe* and the *New York Times* that forces were being marshalled on both sides of the question of leasing Georges Bank for petroleum development. Georges Bank is a rich fishery which begins about 100 miles east of Cape Cod. The oil industry wants the right to look for oil and gas as soon as possible. Not surprisingly, fishermen and environmentalists and other articulate New Englanders and New Yorkers are opposed. The issue was a natural, a confrontation between development and environmental protection advocates. It was complex and emotional and the data were soft and incomplete.

The scientists at Woods Hole Oceanographic Institution, since they live on Cape Cod and study and care about marine resources, took a heartening interest in helping me. Those who provided me with particularly useful information were David Ross, Dean Bumpus, John Farrington, and Howard Sanders. Their role and that of other scientists and officials, it must be emphasized, was to supply me with information. All conclusions and generalizations are solely my responsibility. John Schlee and other employees of the U.S. Geological Survey, both at Woods Hole and Washington, D.C., were also kind in sharing their expert knowledge with me.

Three years ago I would have had neither the ability nor the courage to approach this problem. For giving me the tools and approaches and instincts and confidence to try to solve public policy problems I am especially indebted to Howard Raiffa, Richard Zeckhauser, Michael Spence, Thomas Schelling, Frederick Mosteller, Will Fairly, Richard Neustadt, Graham Allison, David Mundel, and William Capron. After me, the man most responsible for this analysis is Henry Jacoby, who as a practiced environment-energy analyst gave me invaluable encouragement, information, and advice. Three years ago this would have been a less pleasant and fruitful task as I was not married to my wife Sylvia, who aided my attempt to write clearly and to make sense.

William R. Ahern, Jr.
Cambridge, Massachusetts

Oil and the Outer Coastal Shelf: The Georges Bank Case

Chapter One

The Decision Problem
and Possible Outcomes

" . . . wouldn't your case be more persuasive if you could give us
some kind of a, and excuse the expression, cost-benefit formula for
this particular field where we knew what might be given up in the
way of Government and general economic revenues vis-a-vis what
would be gained in the way of esthetic values, and so forth?"

> —Congressman Craig Hosmer to Marvin Levine,
> deputy counsel of Santa Barbara County, at
> 1970 hearings on bills to terminate oil leases in
> the Santa Barbara Channel.

The last major source of moderate cost oil and gas in the United States
is the Outer Continental Shelf (OCS). By far the biggest portion of the continental
shelf, the OCS extends from coastal state seabed boundaries, usually three miles
from shore, out to the edge of the shelf, generally considered to be a depth of
200 meters. The Outer Continental Shelf Lands Act of 1953 gave the Secretary of
the Interior authority to manage the federally owned OCS and to sell mineral
leases on it through competitive bidding. Millions of acres have been leased on the
Gulf of Mexico OCS, primarily off Louisiana and Texas, since 1954. A 1968 sale
leased some of the Santa Barbara Channel. There have been unsuccessful lease
sales, where no commercial petroleum deposits were found, off Oregon, Washington, Northern California, and the Gulf Coast of Florida. To date no federal
lease sales on the Atlantic or Alaskan OCS have been made.

There appear to be three possibly favorable petroleum provinces
under the continental shelf off the East Coast of the United States. Oil companies and the Interior Department have identified these by interpretation of

geological and geophysical data wihch have been collected since the early 1960s. The three favorable areas are the Blake Plateau trough off Florida and Georgia, the Baltimore Canyon trough off the coast from New Jersey to North Carolina, and Georges Bank, southeast of Cape Cod.[1]

The Interior Department cannot hold a lease sale of any of these areas because the question of who holds sovereign rights to the resources under the Atlantic Continental Shelf, the federal government or the adjacent coastal states, is in litigation before the Supreme Court (see Chapter Six). A decision is not expected until 1975 or later. The history of similar cases involving the Gulf states and California indicates that the decision will most likely be in favor of federal ownership outside the three-miles limit. Should the Court rule in favor of the states, however, the decisions to lease the Atlantic Shelf areas would be up to them, and any leasing and royalty revenue would accrue to them. With respect to Georges Bank, the leasing prerogative would belong to the Commonwealth of Massachusetts. In this book it is assumed the decision gives the federal government jurisdiction over Georges Bank.

This book concerns only the problem of whether or not to offer leases for petroleum development on Georges Bank. An orientation map of the Bank, its relation to land and to other features of the ocean floor, is presented as Figure 1-1. The data used here are commonly available for all OCS areas. If the results from this analysis were obtained for each remaining favorable off-shore petroleum province, the areas could be ranked in terms of economic advantages and environmental disadvantages, thus providing a rationale for accelerating leasing of some areas and delaying or preventing leasing of others.

After the Supreme Court decision there will be strong political pressure for leasing Atlantic OCS areas. President Nixon, in his 1971 Clean Energy Message, ordered a doubling of the rate of leasing in the Gulf of Mexico.[2] He also called for leasing in other OCS areas besides those already partly leased. In response, Interior tentatively scheduled a sale of Atlantic leases to be held before 1976.[3] The schedule has slipped, of course, due to the litigation. Then in April of 1973 the President ordered Interior to take steps to triple, by 1979, the annual acreage leased on the federally owned OCS.[4] Reasons for these orders are clear. To make up for declining onshore petroleum production and to prevent reliance on massive amounts of oil imports, the President wants to encourage as much United States production as possible.

Georges Bank, however, is already being used. It is a rich international fishery, the main resource of the non-coastal New England fishing industry. Fishermen perceive that petroleum development on the Bank would hinder their business. Many coastal residents, environmentalists, seashore enthusiasts, and state and local officials fear the possibility of large oil spills damaging their beaches. Articulate New Englanders and New Yorkers feel the risks of environmental damage far outweigh any advantages from offshore petroleum development. Elected officials from the Northeast are responding to deeply felt concerns when they op-

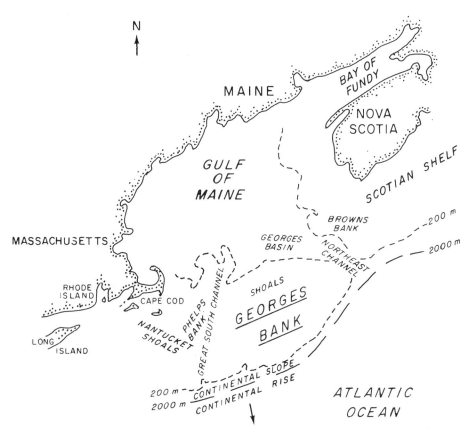

Figure 1-1 Georges Bank Orientation Map

pose offshore oil, as many have done already. Thus, the stage is set for a classic confrontation between pressures for domestic sources of energy and pressures for protection of the environment.

This policy analysis is intended to be as helpful as possible to the Secretary of the Interior and others involved and interested in the decision whether or not to lease Georges Bank. Such analysis should have a number of general characteristics: comprehensiveness, treatment of net instead of gross effects, explicit treatment of uncertainties and the assumptions used to deal with them, and a presentation format enabling easy comparison of advantages and disadvantages of the alternatives.

A comprehensive analysis attempts to predict all major possible outcomes—in this case outcomes such as amount of oil and gas produced, jobs in New England, number and sizes of possible oil spills, and so forth. The potential size of such outcomes should be measured quantitatively whenever possible, since the values of numbers are easily understood and compared. One policy may cause less damage than another. Numbers are simply best for communicating how much less.

The outcomes of the decision would not occur in isolation. Many factors affect the health of an eco-system such as Georges Bank. Other sources in addition to offshore oil production would present the risk of oil spills to the New England Coast. To speak only of the impact of a punch in the nose on the health of a person being hanged may not convey an accurate impression of the actual impact of the punch. Helpful analysis should explicitly discuss the *net* impact of decision outcomes on each valued resource. Thus, many outcomes here are presented in terms of incremental changes instead of in misleading isolation.

It must be emphasized that this analysis of the possible outcomes of a policy decision is a crystal ball activity; it is a prediction of the future. Such a prediction must be based on past experience, on past data. Obviously, many conditions may change, unknown factors may intervene, making such data less applicable to the future. The analysis should present the uncertainties explicitly, that is, it should give the full range of possibilities for each outcome. The range may be extremely wide, especially where data is limited as in estimating how much oil and gas might be underground.

By itself, presentation of a wide range of possibilities is not very helpful. Decision-makers need some impression of the probability of different outcomes. They especially need to know what is "most likely" to happen. In environmental impact analysis much attention is often devoted to the predictions of the "worst credible" risk of damage. Decision-makers need analysis of the size of this and also its probability. In development-environment issues the "worst credible" damage often is considered as if it is the "most likely"

In order to calculate estimates of sizes and probabilities of outcomes when historical data is limited, a number of assumptions are needed. These assumptions should be explicitly laid out so that other analysts can change them to see how such changes might affect predicted outcomes. This format also permits the use of improved data as they become available.

In the analysis of development decisions the costs and benefits of a "yes" decision, a decision to develop, often stand by themselves. The costs and benefits are weighed and the decision is made accordingly. But the decision is really a choice between two alternative packages of advantages and disadvantages, the one of outcomes if the project is undertaken and the other if the *status quo* is not changed. A "no" decision is an active decision with outcomes in the real world. The two packages should be presented side-by-side for the decision-maker to gain an accurate and comprehensive impression of what may result from his actions.

A number of simple descriptive models are used here to break down the outcomes into understandable and doable segments. The first model, the most general, is used to structure the write-up of the analysis.

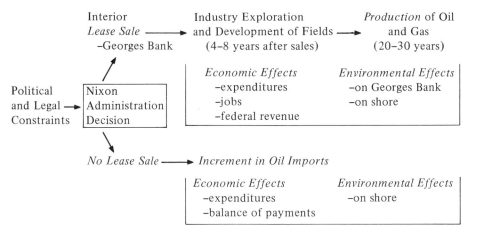

Figure 1-2 General Descriptive Model Linking Decision on Leasing Georges Bank with Possible Outcomes

The Nixon Administration or a subsequent one will be faced with the decision whether or not to offer leases on Georges Bank to the oil industry. The possible general outcomes of the two alternatives are diagrammed in Figure 1-2. If tracts on Georges Bank are leased, industry would require about four to eight years to drill exploratory wells and to establish production platforms and transport facilities, assuming oil and gas are found in commercially producible amounts. The production would last for the life of the fields, perhaps twenty to thirty years. These activities would have impacts in two broad categories, economic and environmental. Industry would make expenditures and create jobs in order to produce the oil and gas and would make payments to the federal government in the form of bonus payments for the leases, in rents, and in royalties on the value of the production. At the same time, if oil is found and produced the risk of oil spills would occur. These large oil spills might damage marine resources on Georges Bank itself, and they might affect the New England shore. Other environmental impacts could result from chronic oil discharges, the presence of oil rigs and other structures on the Bank, and other activities associated with oil and gas production and transportation to shore.

If the decision is made not to lease Georges Bank, the equivalent amount of oil would have to be imported from the Middle East by tanker. This increment in imports would also have economic and environmental effects. Payments of dollars to foreign governments, production inputs, and tanker operators would be required to land the oil at the East Coast. The outflow of dollars would affect the United States balance of payments. As for environmental effects, the increase in tanker traffic needed to import the oil would increase the risk of nearshore oil spills.

This chapter presents a brief introductory discussion of the possible outcomes. Chapter Two deals with the amount of oil and gas production that can be expected from Georges Bank leases. Chapter Three analyzes the economic effects of Georges Bank production and of an increment in imports that would be needed to replace it if not produced. Both general discussions of the environmental impacts of oil pollution and specific discussions of the possible damage to the Georges Bank fishery and the New England shore from oil production and transportation operations are provided in Chapter Four. The major findings and conclusions about these economic and environmental effects of the two alternatives, along with a short set of recommendations, are presented in Chapter Five. The recommendations are based on the predicted possible outcomes in the real world, that is, on expected federal revenues, number of jobs in New England, outflows of dollars, contamination of sea scallops, probabilities of large oil spills reaching certain areas of the New England Coast, and so forth. The presentation of what should be done is then followed, in the last Chapter, by a brief discussion of the legal and political considerations which might affect the decision and its outcomes.

NOTES

1. U.S. Department of the Interior, "Report on Atlantic Coast and Shelf Published; Sent to Eastern States Governors," News Release, Department of the Interior, December 1, 1971.

2. U.S. President, "The President's Message to the Congress, June 4, 1971," *Weekly Compilation of Presidential Documents,* Vol. 7, No. 23, June 7, 1971, pp. 855-66.

3. U.S. Department of the Interior, "Atlantic Outer Continental Shelf (OCS) Leasing," Fact Sheet, Department of the Interior, Office of the Secretary, May 15, 1972.

4. "Energy Message Excerpts," *New York Times,* April 19, 1973, p. 53.

Chapter Two

Oil and Gas under Georges Bank

This chapter addresses four initial questions that must be answered in order to assess the possible effects of leasing Georges Bank for oil and gas development. These are: (1) is there oil and/or gas under Georges Bank; (2) if so, where might it be; (3) how much oil and/or gas might be recoverable; and (4) how much daily production might be expected over the producing life of the area?

PHYSIOGRAPHY AND GEOLOGY[1]

Georges Bank is a northeasterly extension of the Atlantic Coastal Plain and the continental shelf off the East Coast of the United States. (see Figure 1-1) Formerly part of a wide shelf, it is now separated from land by the Gulf of Maine. To the north the Northeast Channel, about 240 meters deep, separates the Bank from the Scotain Shelf off Nova Scotia. To the west the shallower Great South Channel, about eighty meters deep, lies between the Bank and Phelps Bank, Nantucket Shoals, and the rest of the continental shelf to the west and south.

On its seaward side Georges Bank becomes the continental slope at a depth of about 140 meters. The slope descends rapidly to a depth of 2,000 meters where the continental rise begins its gradual descent to the abyss. Shaped like a fat bullet, the Bank is about 200 miles long and 100 miles wide with an area of approximately 10,000 square miles. This is nearly 2,000 square miles larger than the Commonwealth of Massachusetts.

Two deep trenches run under the southeast half of the Bank. Sediment has filled these trenches over millions of years, resulting in sediment thickness of from 1.5 to 6.5 kilometers. Much of this sediment was deposited during geologic periods which have proved to be favorable to the formation of oil and gas. Consequently it is the outer area of the Bank that is most likely to hold petroleum deposits. This area, at its closest, is at least seventy miles from Nantucket Island and one hundred miles from Cape Cod. See Figure 2-1 for a map of this relationship.

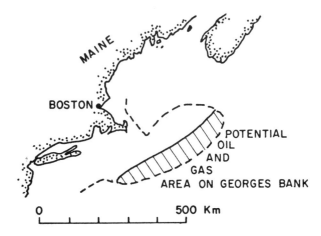

Figure 2-1 Area Potentially Favorable for Oil and Gas Deposits Under Georges Bank

INDUSTRY ACTIVITIES

Two facts suggest that there may be petroleum under Georges Bank. One is that oil companies have drilled about 30 exploratory holes on the continental shelf off Nova Scotia and have found petroleum. The areas of successful finds are similar in geologic structure to Georges Bank. But so far the discovery of commercially exploitable petroleum has been disappointing; the finds have been small. An important aspect of this Canadian experience is that the petroleum discovered has generally not exceeded the kerosene fraction, that is, it has been gas and low boiling point oil.[2,3] The sediments do not seem to contain, on the evidence of drilling so far, the heavy oil fractions which are the major environmental hazard associated with offshore petroleum development. This limited drilling experience off Nova Scotia indicates a strong possibility that drilling on Georges Bank might also discover only small gas fields.

The second indication of a high probability of petroleum under Georges Bank has been the degree of activity by marine geophysical contractors, particularly by Digicon, in the Georges Bank area.[4] Thousands of survey miles have been run over the Bank with the latest and most expensive seismic, gravity, and magnetic sensing methods available. Such extensive and intensive exploration indicates that oil industry geologists and geophysicists have seen favorable indications of possible petroleum bearing structures in the proprietary data gathered by the exploration contractors. This exploration has been conducted all along the shelf off the East Coast. The shelf area seaward of New Jersey to Cape Hatteras seems to present more favorable prospects than Georges Bank.[5]

Table 2-1. McKelvey Estimates of Potential OCS
Oil and Gas Resources

	Total Potential Resources		*Recoverable Resources*	
	oil and NGL[c] billion barrels	*gas trillion cubic feet*	*oil and NGL billion barrels*	*gas trillion cubic feet*
Atlantic OCS	182	423	48	211
Gulf OCS[a]	258	600	69	300
Total OCS[b]	710	1,640	185	820

[a]Includes cumulative production and proved reserves.

[b]Includes Pacific and Alaskan shelves.

[c]Natural gas liquids.

ESTIMATES OF RECOVERABLE PETROLEUM UNDER GEORGES BANK

Geologists cannot look into the ground under Georges Bank to measure the amount of oil and gas there. The director of the Geological Survey observes that, "Satisfactory methods for appraising the magnitude of potential petroleum resources have not yet been developed even for well explored areas."[6] The only way to prove the presence of oil and gas reserves is to drill and find them. Proved reserves are defined as petroleum deposits whose location and magnitude have been confirmed by exploratory drilling and which are recoverable under current economic conditions with available technology. There are no proved reserves on the U.S. Atlantic OCS because there has been no exploratory drilling there. Estimates of potential oil and gas resources have been made by experienced geologists, but their accuracy is highly uncertain. Nevertheless, they are all that are available, so these "best guesses" must be used. Possible resources under the continental slope are not included as they are beyond the reach of foreseeable technology.

The first set of estimates has been made by McKelvey and fellow scientists at the Geological Survey.[7] They show the total volume of favorable sediment under all the U.S. continental shelves and prorate the resulting potential petroleum to offshore regions based on their area and average sediment thickness. Gas amounts are calculated using a gas to oil ratio of 2,500 cubic feet of gas to one barrel of oil. The total potential oil under the shelves is based on extrapolation of the U.S. drilling and discovery experience of 404 billion barrels

Table 2-2.   Nelson and Burke Estimates of Potential OCS
Petroleum Resources

	Liquids *Billions of Barrels*	*Gas* *Trillions of Cubic Feet*
Atlantic OCS	0 - 1.0	0 - 5.0
Gulf OCS	2.0 - 4.0	25 - 50
Total OCS	7.5 - 26	33 - 113

of oil per 1.3 billion feet of exploratory drilling and of the need for one exploratory well for every two square miles of favorable area.

The total U.S. OCS area, out to a depth of 200 meters, is 571,000 square miles. It has an average sediment thickness of 12,000 feet. The Atlantic OCS area covers 122,000 square miles with an average sediment thickness of 12,000 feet. The estimates are made of both the total potential resource in the ground and that which could be recoverable under current economic and technological conditions. These estimates are presented in Table 2-1.

Hendricks makes two sets of estimates using five ratings of areas according to favorability categories. For example, a rating of one, the highest rating, assigns to each 1,000 square miles of an area three billion barrels of oil, eight trillion cubic feet of gas, and 250 million barrels of natural gas liquids (NGL). In his first estimate Hendricks assigns the Atlantic OCS a rating of two, which results in an estimate of total recoverable resources of thirty-five billion barrels of oil and NGL and 183 trillion cubic feet of gas.[8] In his second estimate the Atlantic OCS is given a four rating, resulting in estimated recoverable resources of five and a half billion barrels of oil and thirty-seven trillion cubic feet of gas.[9]

A fourth set of estimates, made by Nelson and Burke, is based on an appraisal of available geologic information on offshore areas, while unknown structures and reservoir formations are not considered.[10] These estimates are presented in Table 2-2.

The last set of estimates, carried out only for the Atlantic OCS, is pro-

Table 2-3.   Five Estimates of Atlantic OCS Potential
Recoverable Petroleum Resources

	Nelson *& Burke*	*Hendricks* *4 rating*	*Spivak &* *Shelburne*	*Hendricks* *2 rating*	*McKelvey*
Oil and NGL (billions barrels)	0 - 1	5.5	6.6	35	48
gas (trillion cubic feet)	0 - 5	37	37	183	211

Table 2-4. Estimates of Potential Recoverable Petroleum under Georges Bank

	Nelson & Burke	Hendricks 4 rating	Spivak & Shelburne	Hendricks 2 rating	McKelvey
oil (billion barrels)	0 - .15	.8	1	5.3	7
gas (trillion cubic feet)	0 - .75	6	6	28	32

vided by Spivak and Shelburne.[11] They estimate Atlantic OCS potentially recoverable resources at 6.6 billion barrels of oil and 37 trillion cubic feet of gas.

The McKelvey and Hendricks estimates can be considered generous because they attempt to take into account the unknowns and uncertainties in a gross fashion. The Nelson and Burke estimate, on the other hand, can be considered highly conservative since it relies only on known geologic information. The five estimates are presented, in order of increasing size, in Table 2-3.

The lowest estimate, zero, is also a possibility. The OCS off Washington and Oregon has been considered a favorable geologic area for petroleum deposits. So far 600,000 acres, about 1,000 square miles, have been leased for $35 million, and about $65 million has been spent on exploration with only small, uneconomical shows of oil and gas resulting from the effort.[12]

The estimates of total Atlantic OCS potential recoverable petroleum resources are highly uncertain, as evidenced by their wide range. Assigning a portion of these estimates to Georges Bank can result in possibly large error because they are made for an entire continental shelf. Such aggregate estimates can encompass wide variations in the local areas of which the entire shelf is composed. Nevertheless, an attempt must be made to measure Georges Bank's possible share of all Atlantic OCS resources. A factor for doing this is the percentage of the

Table 2-5. Estimates of Potential Average Annual Petroleum Production from Georges Bank[a]

	Nelson & Burke	Hendricks 4 rating	Spivak & Shelburne	Hendricks 2 rating	McKelvey
oil (million bbl/yr)	0 - 4	20	25	130	175
gas (billion cu ft/yr)	0 - 19	150	150	700	800

[a]Assumes production period for area of forty years.

Atlantic OCS area represented by Georges Bank. Since the estimates of total Atlantic OCS are to a depth of 200 meters, it is appropriate to use the same limits for Georges Bank. Accordingly, for the purpose of this analysis the Bank, including some adjacent areas, is about 20,000 square miles (10,000 square miles more than the estimate given above), or about fifteen percent of the total Atlantic OCS area of 122,000 square miles.[13] Applying this fifteen percent factor to the estimates of Table 2-3 gives a set of estimates for Georges Bank, presented in Table 2-4.

EXPECTED POSSIBLE PRODUCTION FROM GEORGES BANK

These potential petroleum resources make an impact on the economy and on the environment only when they are produced. Therefore, they must be translated into production figures. The rate of production of such resources would depend on a number of factors, especially on timing of lease sales, the rate of exploration and development by the oil industry lessees, and on characteristics of the reservoirs affecting recovery efficiency.

However, a rough estimate of the possible producing life of the area can be made. Leasing of the entire favorable area on Georges Bank could take five years. If commercial deposits are found, production commonly starts three to five years after the lease is acquired. The average producing life of an identified field is twenty to thirty years.[14] Production from each field increases until all wells are on stream, stays at a chosen recovery rate for a number of years, and then declines. Thus, the reasonable producing life of the area could be forty years, with fields being developed on a staggered schedule. Assuming that all the recoverable resources identified in Table 2-4 would be produced, the average annual production figure can be derived by dividing each estimate by forty years. The results are presented in Table 2-5.

To simplify presentation these five estimates can be reduced to three. The Nelson and Burke estimate is considered to be equivalent to zero when divided and applied to Georges Bank because the four million barrels of oil a year would probably not be worth developing, especially if it is found in a number of geological structures. The Hendricks four rating and the Spivak and Shelburne estimates are averaged to represent a medium estimate and the last two are averaged to represent a high estimate. This reduced set of estimates is presented in Table 2-6.

Petroleum production and consumption data are often presented in daily amounts. Dividing the estimates of Table 2-6 by 365 days gives the daily figures (Table 2-7) which will be useful for future comparisons. It is necessary to test the range of these estimates for realism by comparing them with actual Gulf of

Table 2-6. Reduced Set of Estimates of Potential Average Annual Petroleum Production from Georges Bank

	low	*medium*	*high*
oil (million bbl/yr)	0	23	152
gas (billion cu ft/yr)	0	150	750

Mexico OCS production experience. The potential petroleum resource estimates of Table 2-1 by McKelvey indicate that total Atlantic OCS resources might approximate two-thirds that of the Gulf OCS resources. If Georges Bank resources are assumed to be fifteen percent of the total Atlantic OCS amount of recoverable petroleum, then Georges Bank petroleum might be approximately two-thirds of fifteen percent, or ten percent, of the Gulf OCS potential petroleum resources. Gulf OCS production for 1971 was, after ten years of large scale leasing, 450 million barrels of oil and 3.6 trillion cubic feet of gas. Ten percent of these amounts to 45 million barrels of oil and 360 billion cubic feet of gas, both of which are encompassed in the ranges of Table 2-6.

Another test of these ranges uses data from the most successful lease sale in the Gulf, which resulted in production of 100,000 barrels a day six years after the 1962 sale.[15] This sale offered about 6,000 square miles, and 3,000 square miles were bid for and leased.[16] It is unlikely that operations on the potentially favorable 5,000-10,000 square miles on Georges Bank would be this successful in producing oil, but the 100,000 barrels-a-day estimate does lie within the range of possible daily production from Georges Bank in Table 2-7. This indicates that the high daily oil production estimate, 400,000 barrels a day, is probably extremely high. These two checks show that the production estimates for Georges Bank are of an appropriate order of magnitude and that, given the wide range of possibilities, the estimates seem to be reasonable.

Table 2-7. Estimates of Potential Average Daily Petroleum Production from Georges Bank[a]

	low	*medium*	*high*
oil (thousands of barrels)	0	60	400
gas (millions of cubic feet)	0	400	2,000

[a]Estimates are rounded.

CONCLUSIONS

The analysis of this chapter shows that there is a high likelihood of commercially producible petroleum under Georges Bank. It probably exists under the outer portion of the Bank, at least seventy miles from Nantucket Island and one hundred miles from Cape Cod. The few producible petroleum discoveries off Nova Scotia in sediments similar to those under Georges Bank have been of small gas deposits. There is, then, some chance that there may be little or no oil produced from Georges Bank. Oil production would present the major hazard to the environment. Oil production could be expected to be less than 60,000 barrels a day, although it might, with a very small probability, range up to 400,000 barrels a day over a forty-year production period. Gas production could be 400 million cubic feet a day, a medium estimate, and might range up to 2,000 million cubic feet a day. Confirmation of petroleum reserves must await exploratory drilling.

NOTES

1. Sources for this discussion are: K.O. Emery, "Geology of the Continental Margin off Eastern United States," *Proceedings of the Seventh Symposium of the Colston Research Society* (London: Butterworths Scientific Publications, 1965); K.O. Emery and Elazar Uchupi, "Structure of Georges Bank," *Marine Geology,* 3 (1965), 349-58; Elazar Uchupi, *Atlantic Continental Shelf and Slope of the United States-Physiography.* Geological Survey Professional Paper 529-C (Washington: Government Printing Office, 1968).

2. "Prospects Brighten on Sable Island," *Ocean Industry,* July 1972, p. 51.

3. Interviews with U.S. Geological Survey Scientists, March, 1973.

4. W.H. Luehrman, "The High Cost of Offshore Exploration," *Oceanology,* October, 1971, pp. 24-29.

5. Interviews with U.S. Geological Survey Scientists March, 1973.

6. V.E. McKelvey, F.H. Wang, S.P. Schweinfurth and W.C. Overstreet, "Potential Mineral Resources of the United States Outer Continental Shelf," in U.S. Congress, Senate, Committee on Interior and Insular Affairs, *Outer Continental Shelf Policy Issues, Hearings before the Committee on Interior and Insular Affairs,* Senate, 92nd Cong., 2nd session, 1972, p. 189.

7. *Ibid.,* pp. 166-302.

8. T.A. Hendricks, *Resources of Oil, Gas, and Natural Gas Liquids in the United States and the World,* U.S. Geological Survey Circular 522 (Washington: Government Printing Office, 1965).

9. T.A. Hendricks, source unknown, reported in Department of the Interior, Bureau of Land Management, "Library Research Project, Mid-Atlantic Outer Continental Shelf (Reconnaissance)," (Draft manuscript, Department of the Interior, December 1972), no page numbers.

10. T.W. Nelson and C.A. Burke, "Petroleum Resources of the Conti-

nental Margins of the United States," *Transactions of the Marine Technology Society* (place unknown, 1966), pp. 116-33.

11. J. Spivak and O.B. Shelburne, "Future Hydrocarbon Potential of the Atlantic Coastal Province," *Future Petroleum Provinces of the United States* (American Association of Petroleum Geologists Memoir 15, 1971), pp. 1295-1310.

12. Senate Hearings, *Outer Continental Shelf Policy Issues,* 1972, Part 1, p. 10.

13. U.S. Department of the Interior, *Petroleum and Sulfur on the U.S. Continental Shelf* (Washington: Department of the Interior, December 1969), p. 6.

14. U.S. Department of the Interior, Bureau of Land Management, *Final Environmental Statement, Proposed 1972 Outer Continental Shelf Oil and Gas General Lease Sale Offshore Eastern Louisiana* (Washington: Department of the Interior, June 1972) p. 6.

15. U.S. Congress, House, Committee on Interior and Insular Affairs, *Santa Barbara Channel Leases, California,* Hearings before a subcommittee of the Committee on Interior and Insular Affairs, House of Representatives, on H.R. 18159, 91st Cong., 2nd sess., 1970, p. 80.

16. U.S. Department of the Interior, Geological Survey, *Outer Continental Shelf Statistics* (Washington: Geological Survey, 1972), p. 20.

Chapter Three

Economic Aspects of
Possible Outcomes

FUTURE SOURCES OF OIL AND GAS

The desirability of producing oil and gas on Georges Bank, from an economic standpoint, depends on the role such production might play in the future domestic and international energy situation. Figure 3-1 portrays past and projected domestic sources of energy by source category. The projections are taken primarily from U.S. Bureau of Mines data. They indicate that oil and gas are expected to provide about two-thirds of U.S. energy through the remainder of this century and that their use is predicted approximately to double from 1970 to 1985.

SOURCES OF OIL

Georges Bank oil and gas, if produced, would add to domestic sources of energy. It appears that if it is not produced, the equivalent amount of energy, in the form of oil, would probably be imported from the Middle East. In 1972 United States daily oil imports were 4.7 million barrels, 2.2 million barrels of crude oil and 2.5 million barrels of refined products, primarily residual fuel oil. The imports represented about twenty-eight percent of total domestic oil consumption of 16.7 million barrels. Chase Manhattan Bank reports that total consumption showed a seven percent increase from 1971 to 1972. A six percent increase is forecast from 1972 to 1973.[1]

Domestic oil production, however, decreased by about one percent from 1971 to 1972, remaining at about twelve million barrels a day.[2] Predictions of 1980 domestic production, assuming current or even higher price levels, average about twelve million barrels a day, including projected additions of Alaskan oil.[3,4] This production plateau is because there is no longer any excess domestic production capacity onshore[5] or offshore.[6] The oil industry is meeting increasing difficulty and cost in finding new oil and gas reservoirs large

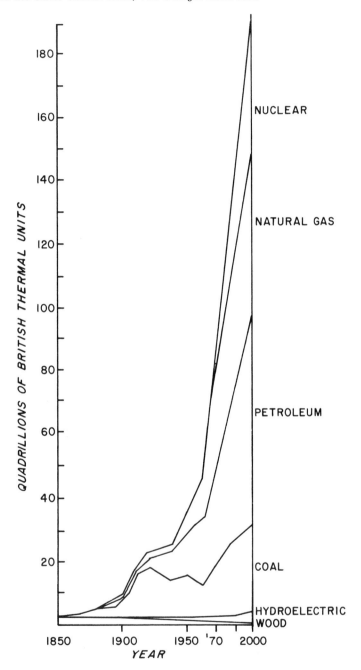

Figure 3-1   Past and Projected Energy Sources. Source: Faltemayer, Edmund, "The Energy 'Joyride' is Over," *Fortune,* September 1972, p. 100. "Bob Weiss Associates for *Fortune* Magazine."

enough to permit economic production. Onshore leasing and drilling efforts have steadily declined during the last decade.[7] At the same time, all OCS production has climbed from about two percent of all domestic oil and gas production to about twelve percent in 1971,[8] and the percentage is expected to climb steadily. The predictions that total U.S. production will remain steady at about twelve million barrels of oil a day through the 1980s reflect inclusion of a constant expansion of OCS production.

Therefore, the federal Oil Policy Committee is reported as being resigned to the fact that oil imports will be needed to supply all—or nearly all—of the increase in U.S. demand for oil.[9] President Nixon has directed the Interior Department to triple by 1979 the OCS acreage leased annually.[10] But this dramatically increased leasing, probably on all U.S. continental shelves, could not significantly decrease the gap between domestic oil production and consumption in the 1980s. Gulf OCS production in 1971 was 1.2 million barrels of oil and ten billion cubic feet of gas a day (the British Thermal Unit equivalent of 800,000 barrels of oil). This level of production was the result of seventeen years of effort, including the drilling of 12,000 wells, the building and placing of 2,000 platforms, the laying of more than 5,000 miles of pipeline, and the investment of billions of dollars.[11] Eventual doubling of Gulf production, from accelerated leasing over the next five years, is already included in the prediction of no increase in total domestic oil production.[12] Thus, Georges Bank production would displace imports no matter what decisions are made about leasing other areas off the Atlantic Coast or other coasts.

A large proportion of these imports will be refined products because U.S. refinery expansion has been small—especially along the East Coast—due to uncertainty about supply of oil imports and to difficulties in finding acceptable sites. The future refinery situation and its impact on the mix of crude oil and refined products imported are beyond the scope of this report. Crude oil and refined products are treated together under the term oil imports.

SOURCES OF GAS

Oil and gas can be substituted for other many industrial, power generating, and commercial uses, and homes can be converted from one to the other for heating. In calorific equivalents, ten billion cubic feet of gas is equal to about 800,000 barrels of oil. Therefore the possibility of gas filling the gap between domestic oil consumption and production must be addressed. Estimates by the Federal Power Commission and others forecast that U.S. gas production will decline from sixty-eight billion cubic feet a day (cfd.) in 1974 to fifty-six billion cfd. in 1980, while demand for clean gas increases steadily. A gap of between eighteen and twenty-four billion cfd. between what potential users would like to buy and actual supply is predicted.[13, 14] Gas prices at the wellhead are regulated by the

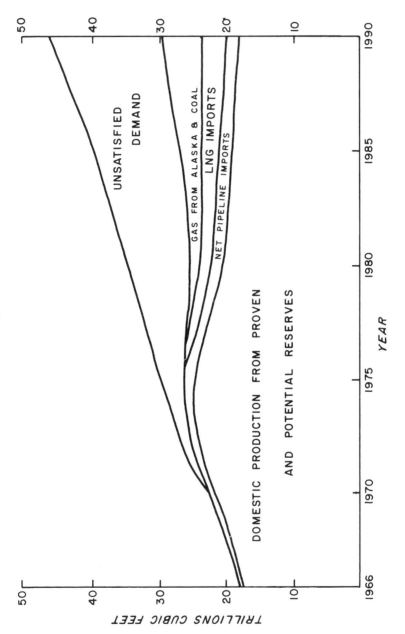

Figure 3-2 U.S. Gas Supply-Demand Balance, 1966–1990. Source: FPC, *National Gas Supply and Demand,* 1972, p. 3.

Federal Power Commission (FPC), and there is general agreement that supply has been, and continues to be, well short of demand at the regulated prices. President Nixon has recently proposed that federal regulation of gas prices be abolished.[15] Should gas prices be permitted to rise, there would probably be more gas domestically produced, but this increase would probably not even be sufficient to meet national demand for gas, let alone to fill the gap between domestic oil production and consumption.

Importing gas is an expensive operation. The natural gas must be liquified and transported in special liquified natural gas (LNG) tankers. To create the capacity to import one million cfd. of gas as LNG is expected to cost about $1.25 million.[16] The equivalent capacity in oil, 800 barrels a day, would cost less than $24,000 investment to import from the Middle East and about $440,000 to produce in the United States (1967 prices).[17] In other words, cost factors are likely to keep gas imports at a low percentage of U.S. energy consumption. This is reflected in FPC predictions of future sources of gas, presented in Figure 3-2. Pipeline imports of gas come primarily from Canada. The predictions in Figure 3-2 show that gas imports from overseas would have to increase at least fourfold to satisfy demand for gas itself. Therefore, it seems safe to conclude that gas imports will not fill the gap between domestic energy production and consumption. On the contrary, oil imports are expected to make up the gap between domestic demand for and production of gas.

EFFECTS OF GEORGES BANK PRODUCTION ON IMPORTS

Oil imports had been regulated by the Mandatory Oil Import Program, administered by the Interior Department and guided by the Oil Policy Committee.[18] Imports were limited to a percentage of domestic demand, and importers needed tickets issued by Interior to bring in a specified amount of oil. The program was intended to minimize U.S. reliance on foreign, possibly unstable, sources of oil and to protect high cost domestic producers from unlimited lower cost imports which would depress prices. However, the price of imports now approaches the most expensive domestic production (to be discussed in the following section) and imports no longer threaten marginal U.S. producers. Because of this and the increasing national demand for oil which cannot be satisfied by domestic production, President Nixon abolished the mandatory oil import quotas on May 1, 1973. In their place will be a protective tariff, termed license fees by the Administration, which starts at $0.105 per barrel of crude oil and $0.52 per barrel of gasoline and which escalates semiannually to November 1975 levels of $0.21 and $0.63 respectively.[19]

At present, about three-quarters of imported oil comes from the Western Hemisphere, primarily from Canada, Venezuela, and the Caribbean. The remaining imports come mainly from West and North Africa, the Middle

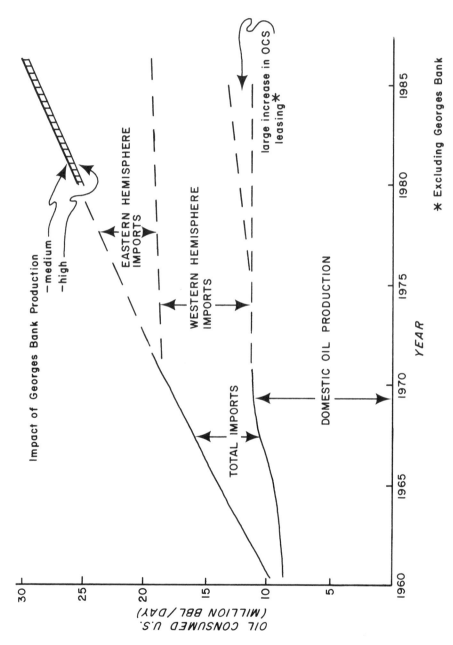

Figure 3-3 Impact of Production of Medium and High Estimates of Potential Georges Bank Oil and Gas on Stream in 1980.

East, and Indonesia.[20] Imports from Canada and other Western Hemisphere sources are considered more reliable, from a national security standpoint, than imports from the Middle East. But the Middle East is the home of roughly sixty percent of total world petroleum reserves while the U.S. share, in comparison, is less than seven percent. Western Hemisphere sources other then the United States have only about seven percent of world reserves.[21] Given the increasing demand for petroleum in these Western Hemisphere nations themselves, and given their limited reserves, it does not appear likely that much of an increase in oil exports to the U.S. would be possible. Canadian exports to the U.S. from western oil fields are only possible because of the large amount of imports brought into eastern Canada, primarily from Venezuela. In early 1973 the U.S. was receiving 1.22 million barrels of oil a day from western Canada via pipeline while one million barrels a day were being imported into eastern Canada–brought there, ironically, in a pipeline from Portland, Maine. In February, 1973 the Canadian government restricted exports to the U.S., disallowing any increase out of fear of shortages to domestic Canadian refineries.[22] Therefore it appears that the increasing amount of oil imports to the U.S. will come from the only area in the world with excess capacity, the Middle East.

The medium Georges Bank daily oil and gas production estimates from Table 2-7 are, together, the total energy equivalent of 92,000 barrels of oil. The high estimates are the equivalent of 560,000 barrels of oil a day. The impact of this potential production on imports from the Middle East, and on the future oil picture in general, is shown in Figure 3-3. The impact would be small. If imports in 1985 are twelve million barrels of oil a day, producing the medium estimate of Georges Bank's potential would decrease total imports by less than one percent. Producing the high estimate would decrease imports by about five percent. Consequently, it appears that Georges Bank oil and gas production would, at the margin, decrease oil imports from the Middle East by the energy-equivalent amount. The same conclusion is reached by the Massachusetts Institute of Technology Offshore Oil Task Group. They report that, "While presently only 6% of U.S. imports emanate from the Persian Gulf, it is expected to increase rapidly. Hence, this is the source on the margin. . . ."[23]

PRICES AND COSTS OF IMPORTED AND GEORGES BANK PETROLEUM

Components of Oil Prices

To assess the size of dollar flows associated with either Georges Bank production or an increment of imported oil requires a comparison of oil prices. Components of the prices of a barrel of crude oil from Libya, Venezuela, Saudi Arabia, and the Louisiana OCS, landed at the East Coast, are presented in

Table 3-1. Components of the Price of a Barrel of Crude Oil Landed at the U.S. East Coast in September 1971 in Dollars

	Libyan high quality low sulfur 40°	Venezuelan 35° medium sulfur	Saudi Arabian 34° medium sulfur	Louisiana OCS low sulfur 38°
production costs	.30	.52	.12	1.53
tax / royalty government take	1.47	.97	1.03[c]	.62
lease cost	.51	.70	.29	.25
	1.98	1.57	1.32	.86
company tax paid cost	2.28	2.19	1.44	2.40
profit, U.S. tax,[b] / other costs	.52	.59	.29	1.34[a]
price f.o.b. producing area	2.80	2.78	1.73	3.74
spot tanker rate	.30	.17	.61	.28
U.S. duty[d]	.11	.11	.11	—
landed price at U.S. East Coast	3.21	3.06	2.45	4.02

Sources: Interior Department, Office of Oil and Gas, *Worldwide Crude Oil Prices* (Washington: Office of Oil and Gas, fall 1971); Bureau of Land Management, *The Role of Petroleum and Natural Gas from the Outer Continental Shelf in the National Supply of Petroleum and Natural Gas* (Washington: Government Printing Office, February, 1972), section II.

[a]Includes $0.12 for transport to shore.

[b]Includes U.S. corporate profits taxes wichh are well below 48% for oil companies due to the depletion allowance, foreign tax writeoffs, and other provisions.

[c]Currently about $1.75 and rising.

[d]To rise to $0.21 by November 1975.

Table 3-1. These are late 1971 data, and changes, mostly increases, have occurred. But the important information here concerns the identity of the parties who receive the various components of these oil prices.[24]

One point that is clear from this comparison is that Middle East oil is cheap to produce. Adelman[25] estimates that the per barrel resource cost of producing oil in Persian Gulf countries is about $0.10, half representing investment costs and half operating costs. He points out that even under a series of the most pessimistic assumptions the cost would no more than double by 1985. Yet the price of a barrel f.o.b. Saudi Arabia is about twelve times the production cost. This is essentially because a cartel of the eleven major oil exporting countries, organized in the Organization of Petroleum Exporting Countries (OPEC), restricts production and colludes on tax rates. These tax rates have been rising dramatically since 1971 and have brought the price of imported oil close to the U.S. price, which currently averages about $3.75.[26] The multinational oil producing and marketing companies have been acting as tax collectors for the OPEC countries.[27] Since the domestic energy sources of the importing countries, especially of Western Europe, Japan, and the United States, are much higher cost sources, the international oil marketing companies can pass on high prices for OPEC oil to the consumers.

Most U.S. domestically produced oil is of moderate cost compared to the very low Middle East production cost. But some U.S. production is expensive, especially from stripper wells which produce only a few barrels a day and from secondary recovery operations. The Oil Import Program, by restricting the amount of low cost imports, has supported a U.S. price high enough to permit these marginal operations to make a profit.[28] It must be emphasized that most OCS oil is not high cost oil. Production cost of Gulf OCS oil runs generally from $1.50 to $1.80 a barrel before royalty and taxes. Georges Bank production costs are expected to be similar. The MIT Offshore Oil Task Group estimates that Georges Bank oil, or its equivalent in gas, could be produced and carried to shore at a cost, before royalty and taxes, of $0.70 a barrel from an extremely large find up to $2.64 for a very small find of 200 million barrels.[29] Interpolating halfway between these cost estimates, which would represent a find of five billion barrels of recoverable oil, gives a cost per barrel estimate of $1.67, in the range of the Gulf OCS production costs. Adding an $0.86 government take for royalty and lease costs gives a price of $2.53 per barrel of Georges Bank oil landed at the U.S. East Coast. Therefore such oil would have been competitive with imported Saudi Arabian crude even at the lower OPEC tax rates of late 1971. Complete elimination of the Oil Import Program, then, would probably not make Georges Bank production uneconomical. It would be a money losing proposition only if the OPEC cartel should be broken and the tax rates decreased to levels lower than those prevailing in 1971. The current trend, on the other hand, is toward rising prices for oil, due both to higher OPEC taxes and inability to expand domestic production. In

Table 3-2.   Comparison of "Who Gets What" from the Buying of a Barrel of Georges Bank Production or Saudi Arabian Crude Landed at the East Coast, Price $3.75 (in dollars)

	Georges Bank barrel	Saudi Arabian barrel
Producer pays to U.S. or Saudi Arabian production inputs (wages, return on capital . . .)	1.47	.12
Pipeline owner gets	.20	
Tanker owner gets		.61
Production and transport payments	1.67	.73
Marketing company pays Saudi Arabia in royalty and taxes (up $1 from 1971 taxes)		2.32
Producing company pays U.S. royalty (1/6 x $3.75)	.63	
Prorated lease cost to U.S. (up to $0.50 due to higher oil prices, more bidding)	.50	
U.S. duty		.21
Total tax paid cost[a]	2.80	3.26
Return to production company, producer and importer same company	.95	.49[b]
Corporate profit tax to U.S. (Assume rate of 20%)	.19	.10
Net company return	.76	.39
Buyer pays	3.75	3.75

Source: Table 3-1.

[a]The total cost to get the barrel of oil to the U.S. East Coast buyer, before U.S. corporate profits taxes.

[b]If producer overseas and U.S. importer are different companies, this return is roughly divided between them.

recent months the price of foreign oil has often exceeded the cost of the high marginal cost domestic producers.[30]

Immediate Dollar Flows Resulting from Importing a Barrel of Oil from the Middle East or Producing a Barrel of Oil on Georges Bank

When a buyer, usually a refiner, purchases a barrel of oil at the U.S. price for crude, landed at the East Coast, several parties have received parts of the price. Different parties, people, or groups receive different amounts depending

Table 3-3. Comparison of "Who Gets What" in Importing from Saudi Arabia or in Producing from Georges Bank Thirty-Five Million Barrels of Oil a Year (in millions of dollars)

Recipients	Georges Bank production	Saudi Arabian crude
Production inputs get	50	4
Pipeline operator gets	7	
Tanker operator gets		21
Saudi Arabian government gets		80
U.S. government gets	46	12
Petroleum companies get net return of	27	13
Buyer pays	130	130

Source: Table 3-2

on whether that barrel of oil has been domestically produced or imported from abroad. A comparison of "who might receive what" from either importing a barrel of oil from Saudi Arabia or producing it from Georges Bank is presented in Table 3-2. Components of the assumed price, $3.75 per barrel, have been changed to reflect current trends. Saudi Arabian taxes are increased by $1.00 to $2.32, and the tariff on imports becomes $0.21 per barrel.

Table 3-3 shows how much of the $3.75 per barrel price would be received by different parties over a year of producing or importing 92,000 barrels a day, about 35 million barrels of oil, the medium estimate of potential Georges Bank production of oil and gas. The estimates of Tables 3-2 and 3-3 are approximate and are based on currently available data. They may already be outdated due to changed OPEC policies, changes in federal policies, or other changed conditions. Their applicability for the future, into the 1980s, can hardly be guaranteed.

These figures can be used, however, to make some general conclusions about which parties receive what amounts of dollars under the two possible outcomes of the decision on leasing Georges Bank. If Georges Bank is leased, the U.S. federal government could receive $46 million a year. Actually it would be be more than that since the lease payments would come as a lump sum at the initial lease sales, not as an annual payment. If banked, such dollars would receive interest, making them worth more in later years. Providers of inputs to production, workers, investors, supply men, and others, could receive about $50 million a year. And the pipeline operator could receive $7 million if the medium estimate of Georges Bank petroleum production is produced.

On the other hand, if the same amount of oil is imported from Saudi Arabia, that country's government could receive most of the dollars paid out by the buyer, $80 million. The tanker operator could receive $21 million, providers of inputs to production only $4 million, and the U.S. federal government only $12 million. The oil industry as a whole could receive nearly twice as much net return from Georges Bank production as from imports, but both are clearly profitable operations. The Georges Bank producer, the international marketing company, and the importer to the U.S. can easily be the same company, one of the seven giant multinational oil companies—Jersey Standard (a large producer in Saudi Arabia), Royal Dutch/Shell, Mobil, Texaco, Gulf, Standard of California, and British Petroleum. These majors could be expected to develop the majority of Georges Bank leases or to import most of the increment of oil from the Middle East. Small oil companies have neither the expertise, the equipment, nor the financial resources to engage in OCS bidding for favorable leases and in OCS production.[31] So the predominant real difference in dollar payments would be between the U.S. government and U.S. workers and suppliers and investors in the case of Georges Bank production, and the Saudi Arabian or other Middle East government and tanker operators in the case of imports.

BALANCE OF PAYMENTS CONSIDERATIONS

The previous analysis has looked at the immediate dollar payments associated with producing or importing 92,000 barrels of oil a day. It has not, however, measured the impacts of the importation on the U.S. balance of payments. This is a complicated estimation process, and results can be rapidly overtaken by events. An attempt to estimate the potential balance of payments impact of an additional increment of imported oil was made by the Cabinet Task Force on Oil Import Control in 1970.[32] The Task Force identifies the components of the money outflow as (1) the "one-time" capital outflow from the U.S. needed for U.S. owned producing companies to add an increment of production in various oil exporting countries; (2) the annual proximate outflow in producing the oil, consisting of tax and royalty payments, capital replacement costs, and local operating expenses; (3) and payments for tanker freight. The assumption is made that the oil would be produced and imported by a U.S. owned company and that the investment in foreign production would come from U.S. dollar sources. The capital outflow is net of purchases of U.S. capital equipment and engineering services. All the cost of tanker freight is treated as a balance of payments outflow. This is because a negligible amount of imports is carried in U.S. flag, U.S. constructed tankers, and because little U.S. capital has been involved in foreign tanker construction, even for foreign flag tankers controlled by subsidiaries of U.S.corporations.[33] Foreign shipyards finance construction up to eighty percent of cost, and many tanker buyers take advantage of this.

　　　　To assess the *net* effect on the balance of payments the return flows of money must be assessed. The Task Force breaks these down into two time

Table 3-4.   Dollar Flows for Hypothetical Incremental Imports of
One Million Barrels per Day ($millions)

	Source Libya	Middle East
1. Net "one-time" capital outflow	−152	−147
2. Annual proximate outflows, net of U.S. company profits	−532	−535
a. tax paid oil cost	(−459)	(−371)
b. tanker freight	(− 73)	(−164)
3. Annual purchases of U.S. goods and services by exporting country	87	108
4. Annual investment in U.S. by exporting country	?	?
5. Cumulative "third-country" return flows after five years		
a. High case, all returned in ten years	200	192
b. Low case, all returned in twenty years	89	85

Source: Cabinet Task Force, *Oil Import Question,* 1970, p. 289.

periods, the first-round return flows and the higher-order effects. The first-round return flows are composed of purchases of U.S. goods and services by the exporting country and investment in the U.S. by that country. The likelihood of additional purchases of U.S. exports can be described as a country's propensity to import from the U.S. For example, for each added dollar that Saudi Arabia receives from exporting oil to the U.S., it spends $0.291 on imports from the U.S., and so that proportion of the outflow returns in the short run. These calculations range from .062 for Algeria through .291 for Saudi Arabia up to .514 for Venezuela. Since much investment by oil exporting countries comes into U.S. stocks, real estate, and other money making opportunities by way of Switzerland, Bermuda, the Bahamas, and the Netherlands Antilles, it is extremely difficult to estimate the size of this short run return flow. After the 1967 Arab-Israeli War the propensity to invest in the U.S. on the part of Middle Eastern countries has declined as they have spent more for military equipment and services. Therefore the Task Force has not made an explicit estimate of the size of this marginal propensity to invest in the U.S. Middle Eastern bankers report that about eighty percent of monetary reserves from oil exports are deposited in U.S. and European banks and about twenty percent is invested in real estate, enterprises, and other money-making opportunities worldwide.[34]

The higher-order return flows occur over a long time period by way of

Table 3-5. Dollar Flows for Hypothetical Incremental Imports of 100,000 Barrels of Oil from the Middle East[a] (in millions of dollars, 1969 data)

1. Net "one-time" capital outflow.	−15	
2. Annual proximate outflows.	−53	
a. tax paid oil cost		−37
b. freight by tanker		−16
3. Annual purchases of U.S. goods and services by exporting country	11[b]	
4. Investment return flow	?	
5. Cumulative "third-country" return flow after five years.		
a. high case, all returned in ten years	19	
b. low case, all returned in twenty years	8	

[a] Similar estimates for importing the equivalent of the high estimate of potential Georges Bank production, about 500,000 barrels, can be derived by multiplying the above figures by five.

[b] 0.29 x 37.

third parties. The short-run flows involve direct, or nearly direct, transactions between the U.S. and the exporting country: these usually happen within a year of the dollar outflow. However, an exporting country may spend its dollars in Europe or Japan or elsewhere. This increases the third party's national income, which in turn gradually induces more imports from the U.S. on the part of the third country, returning $0.95 or more per dollar outflow to the U.S. in this way. But the process may take twenty years. The Task Force calculates, for example, that after five years the cumulative return flow from Saudi Arabia would be $0.425, up $0.134 from the $0.291 first-order return from a dollar spent in that country.

Table 3-4 presents the Task Force estimates of the dollar flows involved in importing a hypothetical addition of one million barrels of oil a day, using 1969 data. Devaluations of the dollar, OPEC success in raising taxes, and other changed conditions over the last four years make these estimates ancient history. They can be used to indicate the general magnitude of the capital outflow and the annual balance of payments impact that could result should Georges Bank not be developed. The medium estimate of potential Georges Bank production is, rounded off, the equivalent of 100,000 barrels of oil a year. Therefore, the dollar flows associated with such an increment in imports can be derived by dividing the estimates of Table 3-4, which hypothesizes an increment of one million barrels, by ten. The result is Table 3-5.

Assuming a one-time investment return flow of two million dollars and a linearly decreasing return of the "third-country" return flow over a period of twenty years, the long run impact of these net outflows on the balance of payments are diagrammed in Figure 3-4.

The outflow and return flows associated with the capital investment would occur only once. Figure 3-4b shows that the annual outflow to buy the imported oil and the long run return flows would occur again each year. The net outflow would be $32 million the first year from this source, but in subsequent years, as return flows from previous years accumulate, the net outflow would decrease in size. Therefore, the impact on the balance of payments would be largest for the first five or ten years after the investment in capacity to import 100,000 barrels a year. Net impact the year of investment would be an $8 million outflow, the first year of importing a $31.2 (32 − 0.8) million outflow, the second year a $28 million (32 − 4) outflow, with steadily decreasing net outflows for twenty years or more until return flows equal $32 million a year (the area under the twenty year triangle of Figure 3-4b). At that time, if the investment is still producing, the impact on the balance of payments would be neglibible. These declining net outflows are portrayed in Figure 3-4c.

By the time Georges Bank oil and gas is in production the oil imported from the Middle East by all industrial countries will have increased rapidly,[35] so the marginal propensity to import goods and services would probably decrease as the exporting countries become satiated. The countries with the most world oil reserves, such as Saudi Arabia and Kuwait, have small populations and cannot absorb imports without limit. Therefore the estimates of return flows in Table 3-5 would most likely be too high if used to predict the impact of the last increment of U.S. imports from the Middle East in 1980s. Adelman projects conservatively that U.S. oil imports from the Persian Gulf could be 10 million barrels a day in 1980 and 15 million in 1985.[36] The marginal increment in imports that might not be displaced by Georges Bank production would be at the tail end of a massive amount of oil imports. The dollars derived by the Middle East country from exporting the last 100,000 barrels to the U.S. might well be used to speculate in world money markets and to support foreign policy objectives at variance with those of the United States.[37]

By themselves, these estimates of net dollar outflows of $32 million or less do not seem large. But if the annual deficit in the U.S. balance of trade in fuels in the 1980s is $20 or $30 billion, a likely possibility,[38] any marginal increase becomes more serious in terms of the need to generate exports in other sectors of the economy, in terms of larger unpredictable holdings of dollars by Middle East governments, and in terms of pressure to devalue the dollar. In general, it can be said that the larger the balance of payments deficit and the more oil imported from the Middle East, the more disadvantageous the increment of imports needed in place of potential Georges Bank production.

Two possible scenarios might drastically change the size of these incremental dollar flows associated with importing 100,000 barrels of oil a day from

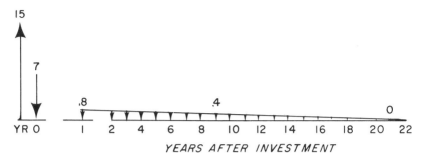

a. Net "one-time" capital outflow ↑ and return flows ↓.

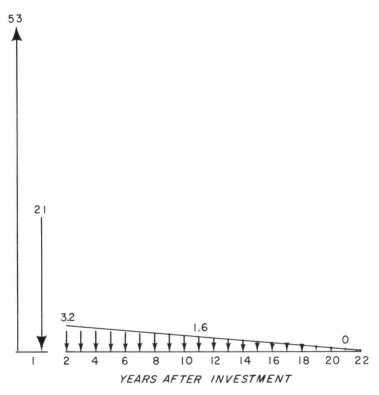

b. Proximate outflows from importing 100,000 barrels the first year after investment ↑ and return flows ↓.

Figure 3-4 Long Run Impact on U.S. Balance of Payments From Investing in a **100,000** Barrel a Day Production Capacity in the Middle East and from Importing **100,000** Barrels a Day for the First Year and for 22 Years After the Investment (in millions of dollars per year).

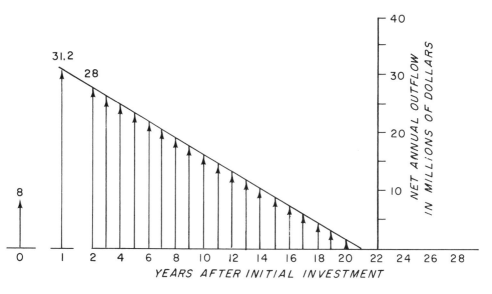

NOTE: ASSUMES LINEAR DECREASING FLOWS

    c. Annual *net* balance of payments flows associated with investing in and importing 100,000 barrels of oil a day from the Middle East (in millions of dollars, numbers rounded).

the Middle East. One scenario hypothesizes that the OPEC cartel would be broken and that perhaps large buyers would make deals with single exporting countries for lower prices in exchange for long run guarantees of giant orders for oil. Perhaps the multinational marketing companies would be removed from marketing, from passing on OPEC taxes to consumers, and relegated to production where they would have to compete for big buyers. Since Middle East prices are fifteen to thirty times production costs, and since large amounts of excess capacity exist there, the cartel might be broken by concerted action on the part of the large oil importing nations.[39] In such a case the OPEC taxes and royalties might decrease and the tax paid cost of such oil could plummet. This, in turn, would result in a large decrease in the annual proximate outflow paid for imported oil. Since the U.S. price would probably remain high, protected from an unlimited flow of cheap imports by a rejuvenated Oil Import Program, the reduced outflow would accrue to the U.S. importer who could charge the protected U.S. price while importing the cheaper foreign crude or to the Federal Government if higher tariffs are used to control imports.

      The second scenario, reflecting the current trend of affairs, simply hypothesizes that the OPEC cartel remains strong and continues to be reinforced by the high-consumption nations and by the multinational oil companies. In this case the annual proximate outflow could increase steadily or rapidly. The "Tehran Agreement" between the OPEC countries and the multinational oil companies has provided for increasing taxes on exported oil. U.S. State Department officials are predicting that OPEC oil could soon reach the current U.S. price of $3.75 a barrel and perhaps reach $5 a barrel by 1980.[40] Consequently, the imports necessitated by a decision not to develop or to delay development of Georges Bank in the 1970s could add $30 million—probably twice that amount—to an already large annual deficit in the U.S. balance of payments in fuels in the 1980s.

OTHER DOMESTIC ENERGY SOURCES

The OCS, including Georges Bank, is the last relatively moderate-cost domestic source of low sulfur content energy. The OCS economic production costs of less than $2. a barrel can be compared with a favorable cost estimate of $5. a barrel for oil shale oil. Alternative domestic energy sources that could actually close the gap between national energy consumption and production generally cost much more, especially nuclear energy; have significant environmental drawbacks, such as strip mined coal and high sulfur coal; or are in early stages of development and could not appreciably affect the U.S. energy picture for decades, such as fusion and geothermal power.[41] So unless there are massive cuts in U.S. energy use, unforeseen technological and production breakthroughs, or national decisions to rely exclusively on higher cost domestic energy sources, Georges Bank production would displace a small increment of imported oil from the Middle East possibly through the turn of the century.

FEDERAL REVENUE

In the case of Georges Bank production the federal government would receive a large amount of revenue. The first component of this would be the winning bids for leases offered at the lease sale. These are called bonus payments. The companies bid for these leases, and full payment must be made within thirty days of winning the lease. If bidding is spirited, if companies are strongly motivated to increase reserves even at the expense of profit, the winning bids for the leases may give the government much of the return from eventual production. The MIT Offshore Study Group reports that, "It appears possible through the medium of competitive bidding for the public body controlling an offshore petroleum resource to appropriate the bulk of the difference between the market value of the landed petroleum and the resource cost of landing that petroleum itself."[42]

A study done for the Interior Department estimates the cost of leasehold acquisitions in the Gulf from 1954 to 1964 at $0.27 per equivalent barrel of oil.[43] This would be a low estimate for current lease sales due to inflation, the rising price of petroleum, and the more intense bidding for the remaining U.S. reserves. Assuming that companies have extensive information on what might be found on Georges Bank and assuming a prorated lease acquisition cost of $0.50 per barrel of oil, the bonus payments for leases to all favorable areas on Georges

Bank could bring the federal government about $700 million (40 years times $0.50 per bbl. times 35 million bbl. per yr), if the medium estimate of Georges Bank recoverable reserves proves true. The total payments would probably be less since little oil is expected to be found. But if the high estimate of reserves is anticipated by the bidders, the bonus payments could range up to $4 billion. Such income to the Treasury, instead of to a Middle Eastern country, could be expected to be valued highly by the President, by the Treasury, and by other officials and agencies. The payments represent income from the sale of rights to federally owned natural resources, and they are not direct burdens on the taxpayers.

The bonus payments would be received whether commercial finds of oil and gas are developed or not. Small rents are collected on the leased acreage until either production starts or the leases are terminated. As soon as production starts, the companies are required to pay royalties of one-sixth the value of the production. Such payments are spread out over the productive life of the lease. At $0.60 per barrel, annual royalty payments from production of the medium estimate of Georges Bank potential would be $22 million—a billion dollars over forty years.[44] Again, it must be emphasized that the royalty payments could well be much lower, or they could range up to $126 million a year given the high production possibility. Such royalties would not represent increased prices to the consumer because the price of oil will probably be well above the Georges Bank cost, including such payments. Therefore the royalty represents a transfer from oil company returns to the U.S. Treasury.

If imported oil substitutes for Georges Bank production the federal government will receive $0.21 per barrel if the 1975 tariff rate remains unchanged. Of course, if the tariff is increased the revenue will be greater. A summary comparison of federal revenue under the two outcomes is presented in Table 3-6.

IMPACTS ON THE NEW ENGLAND REGION ECONOMY

The impacts on the national economy of either developing Georges Bank or importing an increment of oil from the Middle East have been treated as aggregate dollar flows to different parties. The advantages to the federal government from Georges Bank development have been evident in the preceding sections. However, the environmental risks associated with such development would be localized to New England. Therefore, an analysis of the economic impacts on New England is useful for comparing advantages and disadvantages that might result from petroleum development 100 miles off Cape Cod.

Approximate impacts may be derived by an examination of the possible components of the cost of a barrel of oil from Georges Bank. These components[45] are finding costs, producing costs, and transportation costs.

Table 3-6. **Potential Federal Revenues from Georges Bank Petroleum Development and from Importing an Energy Equivalent Amount of Oil[a] (millions of dollars)**

| | Georges Bank Production Estimates | | |
	low	medium	high
Payments for Georges Bank leases, winning competitive bids.	?	700	4,000
Annual royalty payments for Georges Bank production.	0	22	126
Total annual tariff payments on imports.[b]	0	8	42

[a] Does not include corporate profits taxes.

[b] Assumes tariff of $0.21 per barrel.

Finding Costs

Finding costs cover drilling and equipping exploration and development wells, lease acquisition, and geophysical and geologic exploration. Estimation requires prorating the costs for dry holes, overhead, and other joint expenses. The average finding cost in the Gulf OCS has been estimated at $1.19 per barrel.[46] More than forty percent of this is for drilling wells. The drilling of exploratory wells is carried out from mobile drilling rigs which are selfcontained and can be moved from one drilling location to another. They can be submersible barge rigs, self-elevating platform rigs, ship-type rigs, or semisubmersible rigs. Drilling costs may range from $10,000 a day for a mobile rig capable of operating in 50 feet of water to $14,000 a day for one able to drill in 300 feet of water.

Exploratory drilling rigs and crews are generally contracted for by the lease holder. The mobile rigs are owned and built by companies such as The Offshore Company, Odeco, and Global Marine. In the United States, they are located mostly on the Gulf Coast, where offshore oil work began, and in southern California. Thus, any increased construction of mobile rigs, employment in these firms, and equipping would provide purchases and wages mostly to the Gulf regional economy and not to New England.

Most geophysical and geologic exploration has occurred already. It has been contracted to firms based in the South, especially to Houston-based Digicon Corporation. Finally, lease bonus payments accrue to the federal government. As such, a portion would serve to reduce New England federal tax payments, at no noticeable impact.

New England would, however, receive some part of the development costs. The MIT Offshore Oil Task Group estimates that a large find of gas in place—in the range of 5 trillion cubic feet--might require ten platforms distributed in up to five fields. They report that a medium oil find—one billion barrels—might require up to ten platforms in one field.[47]

Once commercial fields have been identified with exploratory drilling, platforms are set up and development wells drilled from them. The cost of platforms depends mainly on the water depth which governs the amount of steel needed to support the drilling rigs and production equipment. A typical platform jacket can cost $2.4 million to stand in 50 feet of water and up to $5.5 million for 400 feet [48] of water.

Lessees prefer that platforms be constructed as close as possible to their eventual location because transportation costs can be high. To bring one from the Gulf Coast to Georges Bank could cost $200,000.[49] Therefore, depending on cost calculations by the lease operators, construction of up to twenty platforms could take place in New England shipyards. On the other hand, the construction might take place in Canada, which is already in the business, or in other East Coast states, particularly if other Atlantic OCS areas are leased before Georges Banks. Currently this seems likely.[50]

The cost of drilling and equipping production wells also varies with water depth and depth to be drilled. Costs can range from a few hundred thousand dollars to a million for very deep wells drilled at angles away from the platform. Drilling crews would be needed, and this could provide employment opportunities for New Englanders. Since operations in the Gulf would also be expanding, at least over the next five to ten years, a large pool of trained workers might not be available for transfer to New England. Thus, some training and employment of available New England workers would occur. Platform crew quarters, oil tanks, flow lines, metering equipment, and other facilities and equipment would be bought. Some of these purchases might be made in New England, although the volume probably would be small.

In summary, the New England share of finding costs would most likely be small. The only large component might be contracts to build up to twenty platforms at two to five million dollars each.

Producing Costs

The average production operating expenses in the Gulf have been estimated at $0.59 per barrel. These consist mainly of expenditures for overhead and for labor and materials to operate the wells and to perform maintenance and workover tasks. The labor and materials expenditures component of the producing costs has been $0.35 per barrel. Since this work would be done either at the platforms or at shore bases over the producing life of the area, most of the expenditure would probably be made in New England. Assuming that all the other producing expenditures go outside the region, the annual spending for labor and materials would be approximately $12 million a year ($ .35 times 35 million barrels a year) for the possible medium estimate of Georges Bank production. It could range up to $70 million a year.

These would be the major annual expenditures affecting the New England economy. Their relative size can be judged in comparison with the total value of the New England catch of fish, $37 million in 1969, and of shellfish, $41 million.[51] In 1971, 225 conventions held in Boston are estimated to have generated about $40 million in direct expenditures there.[52] Such expenditures are "multiplied" in the regional economy as recipients respend some of the dollars in the region. The possible $70 million annual expenditure in New England for Georges Bank production would multiply this way to $90 million or more in the regional economy. The New England region multiplier is about 1.2 or 1.3.[53]

TRANSPORTATION COSTS

Oil can be transported to shore by either pipeline or tanker. Gas, on the other hand, can only be carried by pipeline because technology does not exist to liquify it offshore. Pipeline lengths rarely exceed 100 miles in the Gulf of Mexico, but lengths up to 250 miles may be possible in the near future.[54] Consequently, if an oil pipeline to southeastern New England is not possible because of local opposition, the oil produced on Georges Bank might be carried by tanker to any port from Nova Scotia to Delaware Bay. If the destination is outside New England, few oil transport expenditures would be made in the region.

Oil and gas pipelines to shore terminals on the New England coast could result in some annual expenditures. Pipeline costs are complicated, depending on daily throughput, size of line, line length, and other factors. In addition, rates for oil common carrier pipelines are regulated by the Interstate Commerce Commission and those for gas pipelines by the Federal Power Commission. The MIT Offshore Oil Task Group estimates that oil transport costs from Georges Bank could range from $0.36 per barrel if the find is very small to about $0.04 if it is extremely large. Similarly for gas, costs could range from $0.59 to $0.13 per 1,250 cubic feet, the energy equivalent of a barrel of oil. The costs cited first are for 212 trips a year by a 20,000 ton tanker carrying oil and a six inch diameter pipeline carrying gas. The second set of costs is for one thirty inch diameter oil pipeline and two thirty inch gas lines.[55] Interpolation halfway between these endpoint estimates would represent costs if about five billion barrels of oil is found in place, a number near the high estimate derived in Chapter Two. The costs would then be $0.20 to move a barrel of oil.

The medium estimate of potential Georges Bank gas is considered a large find by the MIT group. Cost of transporting it to shore would be closest to $0.13 per 1,250 cubic feet, perhaps about $0.25 or $0.20 for 1,000 cubic feet. If the reserves discovered are larger than those on which these costs are based then transportation charges would be lower, while if they are smaller, as is more likely, the charges would be higher. The estimates place the annual cost of moving the medium estimate of Georges Bank oil production at about $5 million (23 million barrels times $0.20 per bbl.) and of gas production at about $30 million (150 billion cubic feet times $0.20 per 1,000 cubic feet).

These expenditures would go to pipeline companies who in turn would pay wages and other operating expenses and make the initial capital investment needed to construct the line. Construction of pipe laying equipment and of the pipe would probably occur outside New England. Expenditures for maintenance men and materials and other operating outlays would most likely accrue to the region. Federal Power Commission data indicate that annual operating costs of underwater pipelines can be ten percent of initial cost.[56] Therefore, annual operating expenditures could be approximately ten percent of the annual oil and gas pipeline revenues of $35 million, $3.5 million.

Employment

A set of data on expenditures and employment needed to produce 250,000 barrels of oil a day from fields in the North Sea off Great Britain can be used to estimate potential employment needed for Georges Bank development. These data are presented in Table 3-7. Since the medium estimate of potential production from Georges Bank is the equivalent of about 100,000 barrels a day, and since Table 3-7 shows data for 250,000 barrels a day, application of a factor of two-fifths will provide estimates of operating expenditures and employment for Georges Bank. The annual operating expenditures for this North Sea area would be $31-38 million a year ($625-750 million over 20 years). Two-fifths of that is about $12-15 million, an amount which coincides with the estimate made in the previous section on producing costs. Georges Bank production employment would be 120-160 direct employees. The high estimate of Georges Bank production, 560,000 barrels a day, would be roughly twice the 250,000 barrels a day used for calculation of Table 3-7. Therefore, Georges Bank production employment could range up to 600-800 men. As mentioned above, most of the expenditures and employment associated with exploration and initial development for production and capital investment would be made outside New England.

The potential employment may appear modest, probably about 140 and possibly up to 600. But New England is steadily losing manufacturing jobs, jobs that provide exports for the region or displace imports. The number of employees compares with the 150 jobs lost by the move of the Hampden Brass and Aluminum Company to Tennessee or the 400 jobs lost to Massachusetts by the move of the Allis Chalmers Hyde Park circuit-breaker manufacturing operation to Jackson, Mississippi.[57] The Colonial Candle Company of Cape Cod in Hyannis employs 300 and provides a measure of stability to the area, which depends on seasonal tourists for much of its income. It appears that petroleum production and transportation would create jobs in an area, southeastern New England, which has seen steady declines in manufacturing and fishing employment and which has large numbers of underemployed seasonal workers. Therefore most of these

Table 3-7.   Typical Schedule to Produce Proved Reserves of One
Billion Barrels Recoverable with a 250,000 bbl/day Capacity (90
Million bbl/yr) in the North Sea Off Great Britain

	Exploration (including surveys and exploratory drilling)	Planning/ design (Construction of production wells and transport facilities)	Production buildup	plateau	decline
Time in years	2-6	5-6	3-5	5	8-10
Direct Employ-ment	200-400	1,000-2,000		300-400	
Capital Invest-ment in $ million	25-150	625		125-250	
Operating Expenditures in $ million				625-750	

Source: John L. Kennedy, "North Sea Plans Turned into Tangibles," *Oil and Gas Journal,*
January 8, 1973, pp. 65-69. Courtesy of Shell Oil Company.

jobs could be considered net additions to the regional payroll. The actual number would depend on skills needed, on skills available, and on extent of immigra-
tion.

New England Energy Sources

Consumption of oil in New England in 1985 could be about three mil-
lion barrels a day[58] and of gas about 500 billion cubic feet a day.[59] Therefore, the
medium estimate of possible Georges Bank oil production would be about one-
thirtieth of that and the high estimate about one-sixth. For gas the medium estimate would be three-tenths of New England consumption in 1985, and the high
estimate would be fifty percent greater than total New England use of gas.

However, at present New England does not use crude oil except at a
small refinery at Providence, Rhode Island. The region relies on imports from refineries at Delaware Bay, the Port of New York, and foreign locations, for its petroleum products. Thus, unless some refineries are built in New England Georges
Bank oil would have to be shipped out of the region for processing. Building
new refineries is a difficult enterprise, and none have been built on the U.S. East
Coast since 1959. Those proposed for Maine and Delaware were defeated, primarily
on environmental grounds. In addition, tax breaks, low labor and construction
costs, and other economic factors favor building refineries in the Caribbean and
Canada to supply the East Coast. Construction of new refineries in New England

Table 3-8.   Summary of Possible Economic Impacts on New
England Associated with Potential Petroleum Production
from Georges Bank

Impacts During Production Period	*Production Estimate*		
	low	*medium*	*high*
Annual expenditure for operating	0	$12 million	$70 million
Annual expenditure for transport to shore	0	up to $3.5 million	
Employment during production period	0	120 - 160 up to 600	
Percent of New England oil consumption in 1985	0	3%	17%
Percent of New England gas consumption in 1985	0	30%	150%

Possible contracts to build up to twenty platforms at $2 - $5 million each during development period.

Possible contracts to provide commonly-used materials and support services during
development and production periods.

could have large economic effects on the amount of expenditures and employment associated with offshore production. Such effects are beyond the scope of
this report, which is concerned only with offshore production and transport to
shore.[61]

Georges Bank production of oil would not affect the price of oil in
New England because the U.S. price is set by national interaction of demand and
supply. Expected production would be much too small to affect price.[62] Gas
prices, on the other hand, might be affected. If gas continues to be regulated by
the Federal Power Commission, the price of Georges Bank gas would be set by
that authority. At the current New England price of $0.45-0.50 per thousand
cubic feet there is much unfilled demand for gas. Additional production from
Georges Bank, which landed at shore would probably cost less than this, would
then serve to satisfy some or all of the excess demand and would replace low
sulfur residual oil. If gas prices are deregulated, as President Nixon has proposed, and if they rise sharply, a large find on Georges Bank might serve to
lower prices in the local New England market. Such effects are uncertain at
present.[62]

Summary

The estimates in this section on impacts on the New England regional
economy are predictions based on historical experience in the Gulf of Mexico, on
data for other areas, and on other material of varied applicability to Georges Bank

and New England. Predictions are uncertain about Georges Bank operations because it is not yet known if oil or gas exists there. Moreover, production could not start until at least three or more years after a lease sale, which is well into the future. But the calculations do give a rough impression of the magnitude of possible economic impact of Georges Bank petroleum production on New England. The calculations are summarized in Table 3-8.

NOTES

1. "Oil Demand in U.S. Rose Record 7% in '72, Chase Report Shows," *Wall Street Journal,* February 2, 1973, p. 21.
2. *Ibid.*
3. "The Middle East Squeeze on the Oil Giants," *Business Week,* July 29, 1972, p. 56.
4. See U.S. Cabinet Task Force on Oil Import Control, *The Oil Import Question* (Washington: Government Printing Office, February 1970) for presentations of a number of future scenarios involving domestic oil production.
5. Each month the *New York Times* reports that Texas and Louisiana oil regulatory bodies have set production quotas at 100 percent and still cannot fill demand.
6. Senate *Hearings, Outer Continental Shelf Policy Issues,* 1972, Part I, p. 59.
7. Interior Department, *Final Environmental Statement,* 1972, pp. 153-161.
8. Interior Department, *Outer Shelf Statistics,* 1972, p. 77.
9. Gene Kinney, "Watching Washington," *Oil and Gas Journal,* September 4, 1972, p. 49.
10. "Energy Message Excerpts," *New York Times,* April 19, 1973, p. 53.
11. Interior Department, *Final Environmental Statement,* 1972, p. 7.
12. *Ibid.,* pp. 1-7.
13. Frank J. Gardner, "Vast Worldwide Trade Blooming in LNG," *Oil and Gas Journal,* September 11, 1972, pp. 52-55.
14. U.S. Federal Power Commission, Bureau of Natural Gas, *National Gas Supply and Demand 1971-1990* (Washington: Government Printing Office, February 1972).
15. "Energy Message Excerpts," *New York Times,* April 19, 1973, p. 53.
16. Gardner, "Trade Blooming in LNG," p. 54.
17. Cabinet Task Force, *Oil Import Question,* 1970, p. 278.
18. For a full discussion of the history, regulations, and effects of this program see: Cabinet Task Force, Oil Import Question, 1970.
19. Edward Cowan, "President Offers Policy to Avert an Energy Crisis," *New York Times,* April 19, 1973, p. 53.
20. British Petroleum, *BP Statistical Review,* 1972, p. 10.
21. *Ibid.,* p. 5.

22. "Ottawa's Controls on Crude Exports Fan Friction with Alberta," *Oil and Gas Journal,* February 26, 1973, p. 20.

23. Massachusetts Institute of Technology, Offshore Oil Task Group, *The Georges Bank Petroleum Study,* 2 Vols., Report No. MITSG 73-5 (Cambridge, Mass.: Massachusetts Institute of Technology, February 1, 1973), Vol. 1, p. 42.

24. For extensive discussions on aspects of the domestic and international oil industry see: U.S. Congress, Senate, Committee on the Judiciary, *Governmental Intervention in the Market Place, The Petroleum Industry, Hearings* before a subcommittee of the Committee on the Judiciary, Senate, 91st Cong., 2nd sess., 1970, 5 parts. and M.A. Adelman, *The World Petroleum Market* (Baltimore: Johns Hopkins University Press, 1972).

25. MIT Offshore Group, *Georges Bank Study,* 1973, Vol. 1, p. 43.

26. "Middle East Squeeze," *Business Week,* p. 56.

27. M. Adelman, "Is the Oil Shortage Real Oil Companies as OPEC Tax Collectors," *Foreign Policy,* 10 (December 1972).

28. See Cabinet Task Force, *Oil Import Question,* 1970, p. 217 for the U.S. short run oil supply curve.

29. MIT Offshore Group, *Georges Bank Study,* 1973, Vol. I, p. 114.

30. William D. Smith, "Sun Oil Announces a Rise in Price It Pays for Crude," *New York Times,* March 20, 1973, p. 51.

31. Senate, *The Petroleum Industry, Hearings,* 1970, Part 5, pp. 2005-2006, 2256.

32. Cabinet Task Force, *Oil Import Question,* 1970, Appendix H.

33. *Ibid.,* p. 280.

34. Juan de Onis, "Mastery Over World Oil Supply Shifts to Producing Countries," *New York Times,* April 16, 1973, p. 28.

35. British Petroleum, *BP Statistical Review,* 1971, p. 16.

36. Adelman, "Companies as OPEC Tax Collectors," 1972.

37. For some of these problems see: "The Arab World: Oil, Power, Violence," *Time,* April 2, 1973, pp. 23-31.

38. John G. McLean, "Energy and America," advertisement, *Time,* December 11, 1972, pp. 104-5.

39. There are a few faint rumblings on this subject (U.S. Planning World Organization of Oil-Importing Countries to Cope with Any Shortages," *New York Times,* April 16, 1973, p. 29).

40. A self fullfilling prophecy if they take no action to make the importing countries a monopoly such as OPEC ("Middle East Squeeze," *Business Week,* July 29, 1972, p. 56).

41. For discussions of possible future energy sources, their costs, advantages, and disadvantages, and their prospects of substituting for OCS oil and gas see: Interior Department, *Final Environment Statement,* 1972, Chapter VIII; and Faltermayer, "Energy 'Joyride' Over," *Fortune,* September 1972.

42. MIT Study Group, *Georges Bank Study,* 1973, Vol. 1, p. 157.

43. Interior Department, *Role of Petroleum from the Outer Continental Shelf in National Supply,* 1970, p. 188.

44. Assumes gas prices rise to where the energy equivalent in oil or gas is priced the same.

45. These apply, roughly, to the equivalent amount of gas.

46. Most of this discussion of costs is taken from Interior Department, *Role of Petroleum from the Outer Continental Shelf in National Supply,* 1970, Section 2.

47. MIT Offshore Group, *Georges Bank Study,* 1973, Vol. I, p. 262.

48. *Ibid.,* p. 107.

49. *Ibid.,* p. 173.

50. New England state officials might be able to get some of this business by trading help in getting sites for shore support facilities and pipeline right-of ways across state lands.

51. U.S. Department of Commerce. *Statistical Abstract of the United States. 1971.* (Washington: Government Printing Office, July 1971), p. 623.

52. Massachusetts Department of Commerce, *Commerce Digest,* Vol. 12, No. 3, March 1971, p. 1.

53. John Devanney, MIT Offshore Oil Task Group, oral presentation, MIT, March 22, 1973.

54. National Petroleum Council, *Environmental Conservation-the Oil and Gas Industries,* Vol. II (Washington: National Petroleum Council, 1972), p. 212.

55. MIT Offshore Group, *Georges Bank Study,* 1973, Vol. I, p. 114.

56. Interior Department, *Role of Petroleum from the Outer Continental Shelf in National Supply,* 1970, p. 116.

57. "Shutdowns Keeping Bay State off Balance," *Boston Globe,* March 26, 1972, p. 15.

58. MIT Offshore Oil Task Group oral presentation, Massachusetts Institute of Technology, March 22, 1973.

59. Interior Department, *Role of Petroleum from the Outer Continental Shelf in National Supply,* 1970, p. 116.

60. "Who Shut the Heat Off?" *Time,* February 12, 1973, p. 43.

61. For extensive analysis of the impacts of refineries in New England see: MIT Offshore Group, *Georges Bank Study,* 1973.

62. *Ibid.,* pp. 121-135.

63. *Ibid.,* pp. 136-139.

Chapter Four

Environmental Aspects of Potential Oil and Gas Development on Georges Bank

INTRODUCTION

The future possible impacts on Georges Bank itself and on the New England environment are uncertain because many relevant conditions are unknown or highly variable; even the amount of oil and gas to be produced is unknown. Nevertheless, a large amount of historical data exists on past effects of offshore petroleum development; studies, both in laboratories and after actual accidental spills, have been made on effects of oil on marine life, and broad estimates of possible oil and gas production have been made (above, Chapter Two). This information, coupled with analyses conducted by the MIT Offshore Oil Task Group, permit a general evaluation of the possible adverse, and favorable, environmental impacts of Georges Bank oil and gas development.

Figure 4-1 presents a descriptive model linking Georges Bank petroleum development with various impacts on the environment. The petroleum development is divided into two parts: the production operations which all take place on Georges Bank, and the transportation network which might move oil and gas to shore. These are treated separately because major oil spills far offshore and near or on the shore have significantly different impacts. Actual production and transport of oil are necessary to give rise to a possibility of oil spills. If primarily gas is found, potential environmental damage would be negligible.

The first section of this chapter predicts ranges of possible numbers and sizes of oil spills that might occur on Georges Bank over a forty-year petroleum production period. This is followed by a general discussion of the natural life of oil spilled on water. Different components of oil may evaporate, float, dissolve, and sink under varieties of weather and sea conditions. Over time an oil spill can change character, weathering and being degraded and dispersed until its remnants eventually are broadly scattered and the spill has "disappeared." Over the life of an oil spill the different components of the oil in various parts of the marine environment can damage marine life, with effects ranging from

45

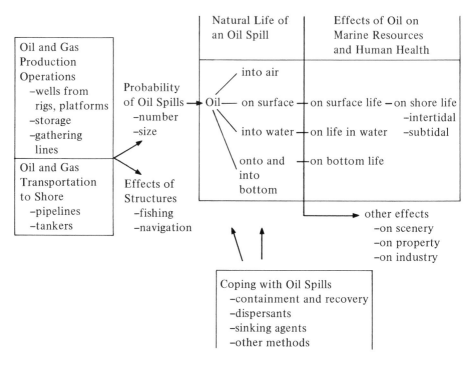

Figure 4-1 General Descriptive Model Linking Georges Bank Oil and Gas Development with Impacts on the Environment

immediate death to subtle disruption of chemical communication between members of a species. Many other human-caused factors and variations in natural features of the marine ecosystems also affect marine life, making identification and measurement of damage caused by oil a difficult task. However, as the MIT Offshore Oil Task Group points out, ". . . considerable information has been amassed on oil pollution and we feel that one can make rather strong statements in certain areas."[1] The gross outline of possible damage caused by oil spills both on Georges Bank and near shore is available even though much detailed information has not been gathered. Speculations have been made that oil in the ocean may cause deleterious effects on human health. But evidence for this is limited at present.

The impacts of oil production operations on the Bank depend on the possible routes taken by potential major oil spills. Therefore, another section deals with predicting the possible number of spills that might reach shore or stay on the Bank for some time or soon travel away from shore and off Georges Bank. Other effects from oil production operations could arise from chronic discharges of oil and wastes and from the presence of structures on the Bank.

Table 4-1. List of Major OCS Oil Spills, 1964-1971 (All Gulf OCS unless otherwise noted)

Dates/Year		Company/Cause/Control Method/Volume
1/20-27	1964	Gulf Oil / lost control of completed well / controlled by bridging / 500 bbl.
4/8-9,	1964	Continental Oil / platform struck by freighter / fire controlled / 2,600 bbl.
10/3,	1964	Signal and Tenneco Oil / loss of oil stored on platforms due to hurricane / 7,000 bbl.
6/21-27,	1965	Continental Oil / well break while abandoning wells / cementing / 500 bbl.
2/5,	1966	Union Oil / blowout during workover / valve installed / 15 minute spill.
1/13	1966	Texaco / oil and gas blowout / ceased on its own / two day spill.
10/15-27,	1967	Humble Pipeline Co. / pipeline break caused by ship dragging anchor in storm / line shut and repaired / 160,000 bbl.
2/21,	1968	Gulf Oil / oil blowout during workover / valves closed / one hour spill.
3/12,	1968	Gulf Oil / pipeline break caused by dragging ship anchor / line repaired / 6,000 bbl.
1/28-2/7,	1969	Union Oil / blowout and formation damage / cemented / 80,000 bbl., much of which seeped during the year (Santa Barbara)
3/16-19,	1969	Mobil Oil / oil blowout / capped / 2,500 bbl.
12/16,	1969	Union Oil / pipeline break / line repaired / 900 bbl. (Santa Barbara)
3/10-31,	1970	Chevron / fire, wells without storm chokes / relief wells / 30,500 bbl.
5/28-30,	1970	Chambers and Kennedy / explosion and fire / storage tank loss of 1,400 bbl.
12/1, 2/1,	1970- 1971	Shell Oil / explosion during wireline work on well / relief wells / 53,000 bbl.
10,	1971	Amoco / fire / 450 bbl.

Sources: U.S. Congress, House, Committee on Appropriations, *Department of the Interior and Related Agencies Appropriations for 1972, Hearings* before a subcommittee of the Committee on Appropriations, House of Representatives, 92nd Cong., 1st sess., 1971, pp. 185-216, 546.

U.S. Department of the Interior, Bureau of Land Management, *Draft Environmental Statement Proposed 1972 Outer Continental Shelf Oil and Gas General Lease Sale Offshore Louisiana* (Washington: Interior Department, July 1972), pp. 71-124, National Petroleum Council, *Environmental Conservation*, 1972, pp. 242-256.

Senate, *Shelf Policy Issues, Hearings*, 1972, p. 309.

Alan A. Allen, *Estimates of Surface Pollution Resulting from Submarine Oil Seeps at Platform A and Coal Oil Point* (General Research Corporation Technical Memorandum 1230, 1969).

Possible impacts from moving oil and gas to shore would depend on the mode used, tankers or pipelines. Such transportation of oil can create the risk of fresh nearshore oil spills. Since the New England coast already is threatened by such spills from the heavy traffic of oil tankers and barges in and out of a number of ports, the possible net changes in the total risk of nearshore spills resulting from possible transport of Georges Bank oil and gas is calculated using oil movement along and to shore as a proxy for this risk.The chapter concludes with a brief discussion of possible measures to cope with oil spills.

In a number of places, quantitative calculations are used to derive numbers of spills, probabilities of spills reaching shore, added risks of oil spills, and other such figures. It must be emphasized that these are predictions based on limited historical data that may well not be applicable to conditions in the future. The numbers are intended to provide general impressions of the size and ranges of possibilities. Each calculation involves a train of assumptions. No claim is made for their precise accuracy, but a claim is made that they appear reasonable and are an improvement over random speculation and ignorance.[2] They also provide a structure for incorporating new data should they become available.

PREDICTING NUMBERS AND SIZES OF POSSIBLE OIL SPILLS FROM GEORGES BANK PETROLEUM OPERATIONS

In order to estimate the possible amounts of environmental damage that Georges Bank oil and gas development might cause, an attempt must be made to predict the possible number of oil spills of different sizes that might occur during the hypothesized forty-year producing life of the area. The focus here is on large oil spills because gas blowouts generally do not cause any environmental damage; the gas vents to the air and dissipates rapidly. The number of OCS oil spills in the future may be different from that in the past, for changes may occur in technology to prevent and control oil spills and in other factors that make spills more or less likely. A list of major OCS spills of 450 barrels of oil or more for the period 1964 through 1971 is presented in Table 4-1. Estimates of the amount of oil spilled in an accident can be highly uncertain, especially for spills from blowing wells whose flow rates are unknown. Controversy still exists over how much was spilled in the 1969 Santa Barbara Channel blowout and subsequent seepage from cracks in the ocean floor. In Table 4-1 a high estimate is cited. Spills from pipelines and storage tanks are more easily measured since the volume stored or carried is generally known. The data prior to 1969 are probably incomplete since there was little attention paid to oil spills and their possible environmental effects. But these are all that are available; they are used in the subsequent analysis. Table 4-2 summarizes the spill data and presents them alongside platform and production information for each year.

The tables show that an OCS oil spill larger than 450 barrels is a rare

**Table 4-2. Major OCS Oil Spills by Year, Number
of Fixed, Producing Structures, Total Number of Wells,
and Annual OCS Oil Production in Barrels**

	Number of oil spills	*Volume of spilled (barrels)*	*Number of producing structures*	*Total number of wells*	*Annual oil production (million bbls.)*
1964	3	500	1,100	6,053	122
		2,600			
		7,000			
1965	1	500	1,200	6,768	145
1966	2	? (15 min.)	1,325	5,702	189
1967	1	160,000	1,450	6,586	222
1968	2	? (small)	1,575	7,575	269
		6,000			
1969	3	80,000	1,675	8,493	313
		2,500			
		900			
1970	3	30,500	1,800	9,392	361
		1,400			
		53,000			
1971	1	450	1,900 (approx.)	10,234	419

Note: Total oil production – 2 billion barrels.
         Average annual production – 250 million barrels.
Source: Interior Department, *Shelf Statistics,* 1971, pp. 27, 75.

event. There were sixteen in eight years, resulting from a variety of causes and from a number of sources. Table 4-3 summarizes the spills by proximate cause.

Three of the spills came from pipelines, two from storage on platforms, and eleven from blowing wells. Of the eleven well spills, four occurred during drilling and three during workover maintenance activities, two were caused by fire on the platform, one by collision with a ship, and one occurred while wells were being abandoned. Given this small number of spills, with the variety of sources and causes and periods in the life of a field during which the accidents occurred, it does not seem possible to predict the number of Georges Bank spills by comparing past variables, such as number of wells or ship traffic or miles of pipeline, with future Georges Bank variables. There are too few

Table 4-3. Major Contributing Causes of the Sixteen
Major OCS Oil Spills (from Table 4-1)

Cause	*Number*
Blowout, loss of well control during drilling	4
Loss of well control during workover	3
Fire	2
Pipeline broke by ship anchor	2
Platform struck by ship	1
Stored oil on platform lost due to hurricane	1
Well break while abandoning wells	1
Pipeline break (cause not known)	1
Explosion on platform, loss of stored oil	1
Total	16

Source: Table 4-1

spills in each category for predictions based on functional relationships between such variables. There is also wide uncertainty about potential Georges Bank development, about number of wells to be drilled, methods of transport to shore, frequency of workover operations, and so forth.

However, one variable has been estimated for Georges Bank, the possible annual production of oil. From Table 2-6 the medium estimate is 23 million barrels a year and the high estimate 152 million barrels. The sixteen spills from past OCS oil production occurred over an eight year period during which oil production averaged about 250 million barrels a year.[3] This comparison is valid only if other important variables that affected past spills and which might affect Georges Bank spills are comparable. Some can be considered comparable: (1) numbers of wells and structures for equal amounts of production; (2) ship traffic; (3) drilling crew training; and (4) potential for human error. But on a number of variables that may contribute to spill accidents, those applicable to Georges Bank operations would result in a higher risk of spills. Some of these are: (1) colder, windier, more severe offshore weather conditions than the Gulf, (2) longer pipeline mileage to shore, and (3) further distance to bring men and equipment to control and stop spills. On the other hand, a series of important variables applicable to future OCS operations would contribute to a lower risk of spills from Georges Bank than from past OCS activities. Some of these are: (1) more stringent federal regulations which require many backup safety devices, use of the latest equipment, frequent testing and training, and other provisions; (2) heightened oil industry sensitivity to environmental concerns; (3) tougher spill liability laws; (4) improved spill prevention and control technology; (5) mandatory burying or trenching of OCS pipelines to prevent contact with anchors; and (6) more stable geologic structures than in the Santa Barbara Channel. The importance of this last set of variables indicates that the expected number of spills from Georges Bank operations would most likely be fewer, for the same amount of oil production, than those resulting from past OCS operations. For simplicity in analysis, though, the assumption is made here that the last two sets of variables—those that might contribute to both more and fewer spills—approximately balance each other. Total oil production, then, is a suitable proxy variable for all the variables.

One more step is needed before a useful prediction of Georges Bank spills can be made. The past spills varied widely in size, and it would be misleading to predict only the number of spills without reference to size. The volume distribution of the sixteen past spills is presented in Figure 4-2. The sizes cluster into three groups: small, 450-10,000 barrels; medium, 10,000-100,000; and large, more than 100,000.

The potential producing life of Georges Bank could be forty years, but historical spill data are available for only eight years. In order to predict forty years of spills the data are assumed to be repeated four more times. In other words, the eight year spill experience during average annual production of 250 million barrels a year is multiplied by five. Assuming that two of the spills of unknown size are in the small category and the other in the medium gives eleven small, four medium, and one large spill on the OCS from 1964 through 1971. A hypothetical forty year period would see, then, fifty-five small, twenty medium, and five large oil spills. The average size of the small spills was 2,400 barrels, the medium 55,000 barrels, and the one large spill was 160,000. This information, with the expected number of spills from Georges Bank operations, is summarized in Table 4-4. The number of spills is derived by using the ratio of possible Georges Bank production to 250 million barrels a year. The medium production estimate gives a ratio of about one-tenth (23/250) and the high three-fifths (152/250).

The expected number of spills on Georges Bank does not indicate how many spills of each size are possible. The range of possible spills is needed to assess risks involved in oil production. That there will be no large spills at all does seem possible, even over forty years. At this writing there have been no major oil spills in the Gulf or off Santa Barbara since the late 1971 Amoco spill of 450 barrels. On the other hand, to estimate the maximum credible environmental risk that might be taken should the decision be made to lease Georges Bank, some impression of the upper range of possible spills, especially of the large spills, must be provided.

The Poisson probability distribution is frequently used to estimate the chances of different numbers of accidents occurring.[4] Applying this distribution, with the expected number of spills from Table 4-4 as the mean parameter, results in the probabilities of different numbers of spills in the three

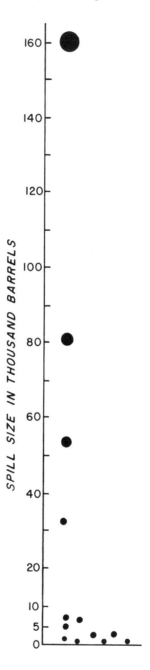

Figure 4-2   Sizes of Thirteen Large OCS Oil Spills (in thousands of barrels).

Table 4-4.   Expected Number of Spills on Georges Bank,
Forty Year Production Period

Spill Size Category	Production Estimate		
	low	medium	high
Small (450-10,000 bbl.) mean size – 2,400 bbl.	0	5.6	33
medium (10,000-100,000 bbl.) mean size – 55,000 bbl.	0	2	12
large (100,000 bbl. plus)	0	0.5	3

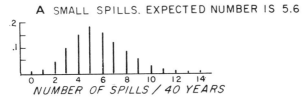

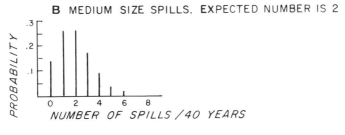

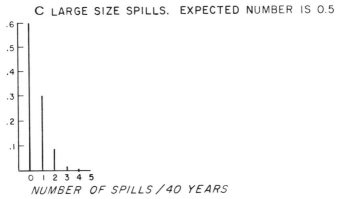

Figure 4-3   Poisson Probability Distributions for Numbers of Oil Spills on Georges Bank Over a Forty Year Production Period (Medium Production Estimate).

Table 4-5. Possible Numbers of Oil Spills from Georges Bank Oil Production over a Period of Forty Years

Spill Size Category	Oil Production Estimates				
	low	medium		high	
		expected number	upper estimate	expected number	upper estimate
Small (less than 10,000 barrels)	0	5.6	14	33	70
Medium (10,000-100,000 bbl.)	0	2	8	12	25
Large (100,000 bbl.+)	0	0.5	4	3	10

size categories for the medium production estimate presented in Figure 4-3. By imposing Poisson on the estimates of possible numbers of spills, the range becomes more apparent. Small spills might be between zero and fourteen over the forty-year production period if the medium production estimate proves to be close. Up to eight medium spills might be expected. There would be a good chance no large spills might occur, although the possibility could exist for up to four. This high number of possible spills, it must be emphasized, is derived from imposing a statistical structure on the possibilities. In the real world, many more spills could be possible, though that would be highly unlikely if the historical experience is at all applicable. The statistical analysis presented here provides a general impression of expected and maximum possible numbers of spills.

Applying the Poisson distribution to the expected numbers of spills from the high Georges Bank oil production estimate also gives a better impression of the extremely remote possibilities. Such analysis indicates that, if the high production estimate is close to reality, up to about seventy small spills might occur during the forty year production period, up to twenty-five medium ones, and up to ten large spills of 100,000 barrels or more. These findings, derived from the assumption that the numbers would fit a Poisson probability distribution, are summarized in Table 4-5.

OCS oil spills seem to exhibit a seasonal pattern. The past OCS oil spills per month are presented in Figure 4-4. Of the sixteen, only three occurred in the spring and summer months while thirteen occurred in the colder months. This pattern is explainable. Weather is rougher in winter, hurricanes occur in the fall, daylight is restricted, and offshore men and equipment are under more stress. All these conditions make accidents more likely. Therefore, it appears that a majority of the possible spills on Georges Bank would probably occur

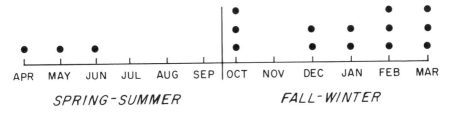

Figure 4-4   Numbers of OCS Oil Spills, 1964-1971, by Month in Which Accident Occurred. Source: Table 4.2.1.

during fall and winter. This observation is relevant to calculations of the number of spills that might reach shore.

Tiny OCS Oil Spills

Many small oil spills occur each year because of both normal OCS operations and accidents. The large spills account by far for the larger proportion of the total amount of oil spilled. Coast Guard data for 1971 show that OCS production operations spilled 2,800 barrels of oil in 1,085 spills, an average of 2.6 barrels per spill. Offshore pipelines more than three miles from shore accounted for 350 barrels in 157 spills, an average 2.2 barrels per spill. Closer pipelines are not cited here because, as in Louisiana, they are old and corroded and serve nearshore operations and those in the coastal marshes.[5] Therefore, such spill data would not be applicable to Georges Bank operations.

This total of 3,150 barrels, spilled in small accidents throughout the year in widely dispersed locations, appears small when compared to the two large spills in the Gulf in 1970 of 30,500 and 53,000 barrels each. Gulf OCS oil production in 1971 was 419 million barrels. Therefore, 0.00075 percent of it was lost through small spills. Applying this factor to the Georges Bank medium production estimate of 23 million barrels a year has small spills amounting to about 170 barrels a year. For the high production estimate the annual total would be about 1,100 barrels.

THE LIFE OF SPILLED OIL

Petroleum Components

Petroleum is a naturally occurring complex mixture of hydrocarbons and other compounds that is not miscible with water.[6] Some components of oil have much more serious biological effects on marine life than others; their behavior, evaporation and solubility, and their amounts in crude oil are highlighted here. Crude oils taken from different locations generally possess the same compounds, but the amount of each can vary widely. The main fractions of oil and their characteristics are presented in Table 4-6.

The fractions that have been found to be toxic to marine life are the

Table 4-6. Characteristics of Fractions of Oil

Fraction	Description	% by weight in crude oil	Density gm/ml	Boiling Point °C	Solubility (gm/10⁶ gm H₂0)
1	Light Paraffins	.1-20	.66-.77	69-230	9.5-.01
2	Heavy Paraffins	0+-10	.77-.78	230-405	.01-.004
3	Light Cycloparaffin	5-30	.75-.9	70-230	55-1
4	Heavy Cycloparaffin	5-30	.9-1	230-405	1-0
*5	Aromatics 1-2 rings	0-5	.88-1.1	80-240	1780-0
*6	Poly-Cyclic Aromatics	0+-5	1.1-1.2	240-400	12.5-0
7	Naphtheno-Aromatics	5-30	.97-1.2	180-400	1-0
8	Residual (includes non-hydro-carbons)	10-70	1-1.1	400-	0

Source: MIT Offshore Group, *Georges Bank Study*, 1973, Vol. II, p. 195.
Key: *indicates the main biologically toxic fractions.

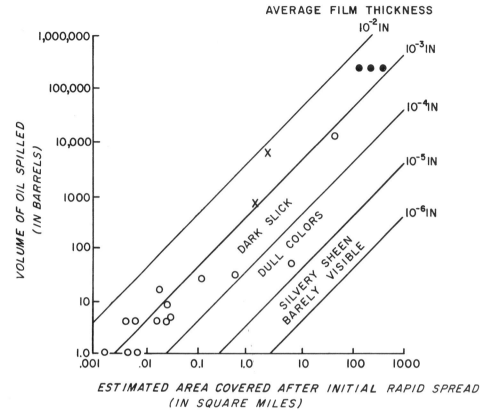

Key: • Observations of Torrey Canyon patches.
      x Calculations.
      ○ Observations from experiments.

Figure 4-5 Volume of Spill, Average Film Thickness After Spreading, and Area Covered. Source: Allan A. Allen and John E. Estes, "Detection and Measurement of Oil Films," in Holmes and DeWitt, eds., *Santa Barbara Oil Symposium,* 1971, p. 65.

aromatics and the poly-cyclic aromatics, as will be discussed in a subsequent section. Fortunately, they represent a small proportion of most crude oils, from little more than zero up to ten percent by weight. Unfortunately, the aromatics exhibit comparatively high solubility in water, especially the one- and two-ring low boiling point aromatics. A fraction is termed low boiling here if its boiling point is below 250°C. The high solubility aromatics are also characterized by low boiling points. These aspects of oil fractions, especially boiling point, density, and solubility, govern the fate of spilled crude oil on water. Three processes affect a spill immediately—spreading, evaporation, and dissolution.

Spreading

Crude oil is less dense than water (density rating 1); therefore, it floats. After being spilled the oil spreads, first because of the gravitational force of the mass of oil and later from the differences in surface tension at the air-water, air-oil, and water-oil interfaces. The spreading is gradually retarded by the force of inertia and the increasing viscous drag between the oil and the water. Hoult calculates that a large spill would rapidly thin out in a number of hours.[7] In general, the more viscous the oil and the lower the water and air temperature, the slower the initial rate of spread and the thicker the slick will be when spreading stops. Calculations, experiments, and observation of slicks indicate that most spills reach a relatively stable average thickness of about .001 inches (.0025 cm) within a few hours to a few days, depending on the size and spill rate of the oil discharge. Average slick thickness and the relationship between spill volume and area eventually covered are shown in Figure 4-5. This figure shows that a spill of 10,000 barrels might cover an area of forty square miles while one of 100,000 barrels could spread over 400 square miles.

Spills with low discharge rates, less than 100 barrels a day, are associated with films thinner than .001 inch. They can spread as thin as .00001 inch, which creates a rainbow effect on the water. Due to the activity of surface-active agents in the oil, mostly compounds with sulfur and oxygen in them, a slick can be surrounded by thin films. These peripheral films can be very large in area yet contain only a small amount of oil. A film one-molecule thick is still visible as a sheen on the water.[8] Consequently, even minute spills of oil can create the appearance of widespread pollution on the water.

Evaporation and Dissolution

Experiments and observations by Smith and MacIntyre show that the low-boiling-point fractions of crude oil on the sea that do not dissolve evaporate almost completely within one to three days, depending on wind speed. As Smith and MacIntyre demonstrate, the faster the wind the faster the rate of evaporation.[9] The low-boiling-point light paraffins and light cycloparaffins nearly all evaporate because they have low solubilities in water. But the one- and two-ring aromatics, the main hazard to marine life, may under some conditions dissolve rather than evaporate. The main condition necessary for these aromatics to dissolve is violent agitation, strong mixing of the surface water and oil by wind and waves. If the oil is immediately mixed with the water, the aromatics may not have a chance to evaporate, and most become dissolved in the water column. Such immediate mixing is necessary for extensive dissolution to occur, because the gas diffusivity rate is one or two orders of magnitude higher than the liquid diffusivity rate. That is, under calmer conditions, from one to ten times more aromatics would evaporate than dissolve because evaporation is faster.

Few poly-cyclic aromatics in crude would evaporate under any conditions because of their high boiling points. In addition, unfortunately, they are moderately soluble in water. Therefore, under conditions of violent agitation much of these toxic compounds would dissolve in the water column. These conclusions are confirmed by Smith and MacIntyre. In samples of spills at sea they found that the low boiling point hydrocarbons predominately evaporated while the greatest proportion of dissolved compounds in the water was the soluble medium molecular weight aromatics.[10]

Role of Particles

Particles in the water can adsorb or absorb oil and carry it into the water column or to the bottom. If particles, particularly clay and organic matter, are present in large amounts at the initial spot of the spill, they can help disperse the oil, still with much of its low-boiling-point content, through the water column if the particles are buoyant or to the bottom if they are heavier. In shallow water under storm conditions the oil may be driven directly into the bottom sediments in addition to being carried there by particles. In such cases much of the lighter oil fractions have no opportunity to evaporate and the entire range of fractions can be incorporated in the bottom sediments.[11]

Quartz sand and gravel bottoms are less likely than clay ones to attach to and retain oil that comes in contact with particles from or on the bottom. The Georges Bank bottom is primarily sand and gravel, while clay bottoms are common close to shore. Therefore, a higher risk of oil persists in nearshore sediments than in bottom areas offshore on Georges Bank. The deeper the water the more dispersed the particles are that sink with oil and the less likely higher concentrations of oil in water are to come in contact with the bottom. Ocean water is generally well mixed to a depth of 100-150 meters. Consequently, rapid mixing of the surface layer of water could bring fairly high concentrations of freshly spilled oil in contact with any portion of Georges Bank, which ranges in depth from four to 140 meters. Obviously, the more shallow the water the higher the concentrations of oil fractions either dissolved or diluted in it.

In summary, two scenarios can be hypothesized. Under calm water and weather conditions, over deep water with a sand or gravel bottom, about ninety percent of the one- and two-ring aromatics will evaporate from an oil spill while the poly-cyclic aromatics will dissolve over a number of days or weeks. Under storm conditions, with high winds and rapid and intense mixing of shallow water, with dense amounts of clay and organic particles in the water, nearly all the one- and two- and polycyclic aromatics could dissolve in the water and become incorporated in bottom sediments where they might persist for years.

Surface Forces

An oil spill does not spread in a perfect circle. Wind and waves break it up into patches and ribbons, the stronger the wind and the more turbulent the water the smaller the patches. These tend to separate over time, increasing the width of the spill's path.[12]

Long Term Weathering

As the slick travels on the water more of its less water-soluble compounds dissolve. Oil that remains in and on the water for any length of time is oxidized and progressively broken down into simpler compounds leading eventually to end products of carbon dioxide and water.[13] Such oxidation can be a highly complex process. Some oil molecules are broken down under ultraviolet attack from sunlight, the photochemical oxidation process, into smaller and more soluble fragments. Others, including the intermediate and higher molecular weight aromatics, are attacked by oxygen through an auto-oxidative process or in the presence of catalysts and enzymes.[14,15] These are very slow processes compared with evaporation and the actions of physical dispersive forces, wind and waves. Oxidation, of course, requires oxygen; on or in the surface water of the ocean this is no problem. Oil at high concentrations in sediments or at chronic high levels in shallow water may deplete an area of oxygen, and such degradation can then take many months or years.

Several species of bacteria attack and decompose and utilize different hydrocarbons. Bacteria first attack the simple molecules, with biodegradability decreasing as molecules become larger and more complex. Such bacteria are fairly common because many hydrocarbons are biogenic, that is, they occur naturally and are synthesized by most living organisms. But the more complex ringed aromatics are not common in biological systems, and only a few uncommon species of bacteria degrade them.[16] In general, bacterial degradation of spilled oil takes place more quickly the greater the number and types of bacteria present, the higher the temperature, the more oxygen available, and the more the hydrocarbons are dispersed. Depending on the state of these and other variables, microbial degradation of oil can take weeks, months, or even years.

As more compounds dissolve and as simple, small molecules are oxidized and degraded, the floating residue of the spill becomes heavier and more viscous. The densities of the heavier oil fractions approach and even exceed that of water (above, Table 4-6). Under some conditions, probably in calm water, the smaller and smaller patches of dense oil residue coalesce into tarry strips or balls. Horn reports finding large numbers of these in surface tows in the Mediterranean and the eastern North Atlantic. They were irregular in shape, 1 mm. to 10 mm. in diameter, and could easily be deformed with the fingers.[17] The long-run fate of these tar pieces seems to be that most gradually sink as they become more dense or weighted down with ecological communities of bacteria and barnacles and other organisms. Some may wash up on shore and some may be ingested by marine life.

Instead of a slick residue becoming tar patches and balls, agitation at the air-sea interface, acting on a relatively thick slick, may eventually form an emulsion with thirty to eighty percent water, which is stable. Such emulsions can be a few centimeters thick and float intact for a number of weeks or months.[18] The tar balls and the stable water in oil emulsions appear to be the fate of patches of large spills that are not dispersed and diluted by the sea in their first days of life. The persistence of tar balls and such emulsions is uncertain, but their lifetimes can be measured at least in months.[19] The significant conclusion here is that a large oil spill on the sea, after a few days—perhaps after a week or more—is not a large, intact mass of "wet" oil. It is depleted of its low boiling point and water soluble fractions, a loss of volume that can range up to fifty percent depending on weather conditions and type of oil. Physical dispersive forces and coalescing of the increasing dense and viscous residue turn the spill into an army of tar balls and strips or a series of patches of oil-in-water emulsion. If the residue remains at sea for a number of months it is further dispersed and chemically and biologically degraded, and residue remnants eventually sink to the bottom of the ocean.

Oil driven into or carried to bottom sediments soon after the spill can be released at a later time by the stirring up or movement of the sediments. Oiled sediment itself may shift with currents and erosion. The deeper the oil is buried the longer degradation can take. Fuel oil buried in sediments at West Falmouth has been shown to persist three and a half years after the spill occurred there in shallow water under storm conditions.[20]

Shore

New England shore areas are basically sand, marsh, or rock. The form in which an oil spill can affect a sandy beach depends on a number of factors, primarily type of oil, age and condition of the slick, type of sand, temperature, and time of day of arrival. Fresh crude oil will penetrate sand, coating sand particles and filling some of the interstitial voids in the beach material.[21] The depth of penetration depends on the amount of oil and the size of the sand particles—the smaller and denser the particles the less the penetration. The oil is deposited between the high and low point to which it is carried on shore, part or all of the intertidal area. Intertidal beach sand is constantly being moved by waves and currents. Thus, over time the oiled sand can be covered over and buried or carried out to deeper water. This movement, the washing and abrasion, and oxidation and biological degradation, will eventually remove the oil from the sand.

Fresh oil reaching a marsh is deposited on the intertidal lengths of the grasses and on the intertidal expanse of mud. Such soft material can soak up

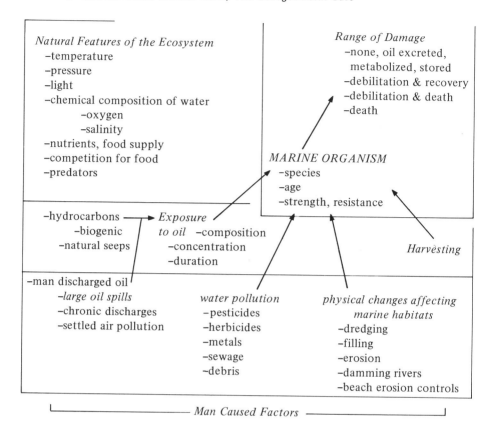

Figure 4-6   Factors Affecting Marine Organisms

the oil. Since marshes are not areas of strong sea action, the oil may persist for a long time. The long-run processes of chemical and biological degradation remove the oil in a matter of weeks or months. Fresh oil can coat rocks over their inter-tidal area and splash up even higher. The oil can be baked dry and hard by the sun, becoming dull colored. Eventually, abrasion by wind and waves wears it away, the more violent the seashore the faster the removal. This can be a lengthy process, taking as long as one to three years for oil above the high tide line.

The impact of an aged spill on shore can be quite different. The tar balls and thick emulsion patches do not sink into the sand but remain on the surface where wave actions and sand abrasion gradually wear them down. The denser the oil the less likely it is to "wet" surfaces although, being viscous, it does stick to relatively flat surfaces. Thus, old oil does not readily coat marsh grasses, and it clings for a long time to rocks only at the extreme high tide mark. Furthermore, since an aged spill would be spread over a wide area in balls and patches, it would be scattered over a longer shoreline than that affected by a fresh spill of equal volume. Local shore areas would not receive high concentrations of the oil.

Conclusion

An oil spill is not a permanent feature of the environment. Some of it evaporates, some dissolves within a few hours. In rough weather it may become rapidly dispersed in water, reaching bottom in shallow areas and remaining in the sediments. Wind and waves break it up into smaller and smaller patches. Floating or dispersed oil is gradually degraded into smaller, different compounds by oxidation and microbial attack. On shore it is deposited in the intertidal area where, eventually, waves and wind wear it away, a process taking days or years.

EFFECTS OF OIL ON MARINE LIFE

Factors Affecting Marine Life

The previous section deals with the physical and chemical effects of spilled oil on air, water, the sea bottom, and shore. Oil also can affect the marine life which occupies these areas, the subject of this section. As with many environmental contaminants, oil pollution from man's activities can have effects that range from the obvious to the subtle. These effects are difficult to identify and measure because so many other factors, generally more significant than oil, also affect marine life. A simplified and general enumeration of those factors is presented in Figure 4-6.

Natural Factors

Every species of marine life requires, for survival, a specific range of temperature, pressure, and chemical composition of the water. Enough food must be available, and the number of predators must not be excessive. Changes in any of these factors—a gradual or sudden increase in water temperature, a decrease in salinity, disappearance of food, a new predator—can wipe out a species in a particular area and, over geologic time, may eliminate the species completely. The natural factors cited in Figure 4-6 can vary widely, and different marine organisms have adapted to the different conditions of the marine environment.[22] These conditions are presented in Figure 4-7.

In general, an organism's ability to withstand changes in natural conditions and to mancaused stresses can be related to where it lives, the habitat to which the species has become adapted. Such general resistance applies to the effects of oil: the more resistant to change and stress an organism is, the less probable the damage done by any concentration of oil. The more severe, the more variable the marine environment is, the more resistant to stress the species living there. In other words, intertidal life is tough. It is exposed to alter-

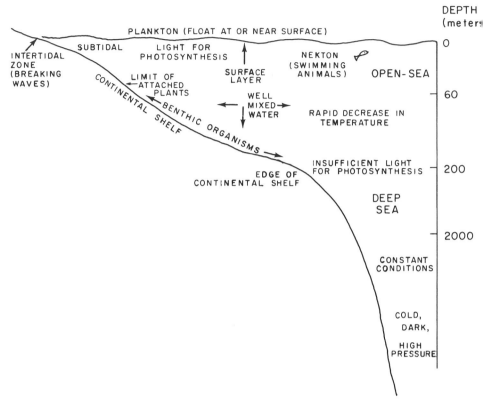

Figure 4-7 General Divisions and Characteristics of the Marine Environment. Source: Ross, *Introduction to Oceanography,* 1970, pp. 141, 147.

nate immersion and exposure to the sun and air. It experiences wide variation in temperature, salinity, and presence of food and is physically buffetted by waves and winds. Many intertidal species are hardy, firmly attached to the substrate, able to close themselves up, protected by shell or chemical substances, and are otherwise well suited to resist most stresses that come their way. They are prolific in producing young since becoming attached to the substrate is a chancy proposition for larvae, and many are swept out to sea. Once attached, however, intertidal organisms are either not mobile, or they are small and cannot travel far. Hence they cannot escape local stresses.

Life in beach sand is sparse because the sand is constantly shifting. Organisms commonly burrow into the sand and are able to close off the burrows to protect themselves, emerging when their habitat is covered by the high tide.[23] The same is typical of life in marsh mud.

Subtidal organisms, on the other hand, are always immersed in water and have not developed the protection and adaptability of intertidal life forms. Non-mobile and low mobility benthic (bottom-dwelling) life—the burrowers, attached organisms, and epi-benthic species—also cannot escape a localized stress, while mobile forms can. As water becomes deeper, variations in temperature, salinity, and other factors decrease. Therefore, life in the sea depths has adapted to narrow ranges of factors and exhibits low tolerance to changes, including increased levels of hydrocarbons in the water.

Nekton—primarily free-swimming fish—can vary considerably in their ability to withstand stress. Fish whose habitat is shallow water are generally strong and adaptable to wide ranges of conditions there, and they generally produce young prolifically. Wide-ranging fish may be viviparous—bearing live young—in which case a litter may be only a few juveniles, but these are more capable of survival than an egg or small larvae. Moreover highly mobile fish can remove themselves from the site of a local stress such as an oil polluted area.

Another category of marine life is plankton—small organisms that float, with no means of self-locomotion, at or near the surface of the water. Plankton may be plant life, phytoplankton, or animal life, zooplankton. The phytoplankton are produced in enormous quantities in the ocean. They convert water and carbon dioxide, through photosynthesis, into organic matter and are the start of the marine food chain. The phytoplankton, like land plants, bloom in spring in temperate waters and decline in the autumn. The plantlife of the marine environment is algae, some of which can be attached to the bottom to a depth of about sixty meters. Some algae float in large forms, known as seaweed, but the vast amount is in the form of single celled planktonic organisms. The phytoplankton live in the euphotic zone, about the top 200 meters of water, where light can penetrate for photosynthesis.

The zooplankton, primarily microscopic animals, graze on the phytoplankton and in turn become food for life forms up the food chain. Many benthic organisms and nekton produce eggs and larvae that live as plankton until they are large enough to swim freely or travel to the bottom. Plankton, especially small larvae, are weak and unable to escape local stresses.

Individual organisms within a species vary widely in their ability to withstand stress. A general observation is that the younger an organism is, the less able it is to fend off dangerous situations and to recover from damage. Naturally, diseased or undernourished marine life also is more susceptible to damage from oil and other challenges.

Two factors affect the amount of exposure marine life has to changed chemical composition of water. The first is feeding habits. Species that filter large amounts of water for feeding come in contact with much larger amounts of oil in water than do predators or grazers. The second is metabolic rate, which can be affected by the time of year. Some marine species have higher metabolic rates in cold water, others in warm water. During the winter there is less life present, and what exists has a slower metabolic rate. Logically, with less movement and a lower demand for food, the organism is less likely to come in contact with oil.

Finally, prior experience with oil, especially with chronic low levels, can affect the sensitivity of life in an area to oil pollution. Over time the life of an area may become resistant to possible damage from even moderate concentrations of oil in water. Such processes are only slightly understood at present.

Human Factors

Oil discharges in the sea are not the only stress placed on marine life by man. Other pollutants, especially pesticides, herbicides, metals, oxygen-consuming sewage, industrial wastes, can seriously affect estuarine and coastal areas. The dredging of harbors, the mining of sand and gravel, the filling of marshes and bays—all these can eliminate crucial spawning and habitat areas for many species of marine life. The damming of rivers can prevent fresh water flow into estuaries, upsetting the salinity variations needed by some species. Many of the activities of a rich and industrialized nation can significantly deplete and damage and change the character of the marine life living near it. A major trauma to a fish or shellfish is getting caught and eaten. Harvesting marine resources can be done intelligently, with regard for the long-run sustainability of fisheries. On the other hand, too much fishing can be the most severe stress a marine ecosystem experiences.

Identifying and Measuring Damage
Attributable to Oil

The short preceding discussion of factors affecting marine life may appear obvious. It is intended to put oil pollution in perspective—as one of many factors, some much more significant, that affect the health of a marine ecosystem. The multiplicity, the variabliity, and the complexity of all these influences, acting in concert, make identification and measurement of the effects of an oil spill, or of chronic discharges of oil, extremely difficult to separate out for identification and measurement. Oysters disappeared from the Gulf of Mexico when oil operations started there in the mid-1940s but research could not link the depletion of the oysters to oil. Moreover, the oysters soon returned and are presently abundant, although oil production there each year is at a new high. One conclusion is that there seem to be large common *natural oyster* fluctuations.[24] Currently, many changes, particularly disappearances, in marine life are attributed to mangenerated pollution. Controversies swirled about the proliferation of the reefeating Crown of Thorns starfish in the South Pacific, about the disappearance of the anchovies off Peru, and about the causes of the fish killing "red tides" off Florida and New England.

It is rarely certain what factors contributed to these events. It is clear, however, that the most serious stress on the Georges Bank fishery is the massive over-fishing of the area by factory ships accompanied by fleets of giant trawlers that can sweep up fish along a fifty-mile-wide line. It seems apparent that many natural and mancaused changes can dwarf by comparison the effects of oil spills in both the short and long-run.

Nevertheless, two methods are available to identify and measure the effects of oil on marine life. One is laboratory experiments; the other, observation of actual accidental spills and their aftermaths.

Laboratory Experiments

Laboratory experiments have a number of inherent limitations in providing predictions of how oil may damage marine life under real world conditions. Oil is introduced into water in which a number of animals of a species are present. The death rate of the animals is measured against the amount of oil introduced into the water. More accurately, the different components of the oil actually in the water are measured. Temperature, salinity, food supply, and the other natural factors are kept constant or eliminated. Precisely because of this control of factors and the simplifications made, the usefulness of laboratory experiment results in generalizing to actual aftermaths of oil spills is limited. But such toxicity experiments do have three valuable functions: (1) identifying the more toxic components of oil; (2) estimating the relative sensitivity of different species to damage by oil; and (3) identifying life stages of species during which increased susceptibility to damage is present.

There now appears to be little doubt about which fractions of oil are the culprits in damaging cells, in contact poisoning of organisms, and in disrupting chemical communication. The MIT Offshore Oil Task Group concludes, "In general, the low boiling aromatic and aromatic derivative components of crude oil have proven to be the most toxic."[25] The other fractions are either toxic at extremely high concentrations, such as the low boiling point paraffins, or have no indicated toxic effects. (All oil components can coat organisms and hinder their movement, smother them and be ingested by them.) Consequently, poisoning effects occur soon after an oil spill when the soluble aromatics can be present in the water under the spill in high concentrations. The soluble aromatics, a small percentage of the spilled oil, are rapidly diluted by mixing with the ocean water. In shallow and enclosed bodies of water, however, the soluble aromatics may attain and remain at high concentrations. Different concentrations can have widely varying effects on different species, as has also been shown with laboratory experiments.[26]

Among the benthic organisms the most sensitive seem to be the crustaceans: lobsters, shrimp, worms, and others that exhibit toxic responses at levels of 1-10 parts of soluble aromatics per million parts of water (ppm). The bivalves— oysters, clams, and mussels which can close their shells when they sense contamination—exhibit moderate resistance; damage can start at 5-50 ppm. soluble aromatics. The gastropods—snails and limpets, that seem able to eat anything—that eat oil and excrete it as feces, are highly resistant to even the soluble aromatics, tolerating concentrations of 100-200 ppm.

Toxicity for finfish starts as low as 0.3 ppm. and can be as high as 50 ppm. depending on the strength and resistance of the species. Mobile fish can, however, recognize very low levels of aromatics in the water and swim away from contaminated areas. This indicates that oil spills on the sea would damage only small numbers of nekton.

Phytoplantkon sensitivity also varies over a wide range, most organisms having tolerances up to 100 ppm. aromatics. On the other hand, cell division for some species can be inhibited by concentrations of 0.1 ppm. The major threat seems to be to the larval stages of benthic organisms and nekton, especially to planktonic larvae. They are ten to 100 times more sensitive to aromatics than adults, being weakened and killed by levels of 0.1-1 ppm. The eggs are somewhat less sensitive to toxic effects since they are protected by the shells.

These concentrations are threshold levels of soluble aromatics at which toxic responses occur in some species of each category of marine life mentioned. Obviously, the higher the concentration above this threshold level, and the longer the duration of exposure, the more serious the toxic effect; eventually the organisms die.[27]

The major toxic effects of oil are (1) direct damaging and killing through coating and asphyxiation, (2) direct kill through contact poisoning of organisms, and (3) damaging and killing through exposure of organisms to the water-soluble aromatic components of the oil. Another mechanism by which oil and other substances may hamper marine life is the disruption of chemical communication between members of a species, usually species that are behaviorally complex and less resistant to stress. Studies suggest that some marine organisms rely on chemical communication for basic life processes such as mating and feeding. This communication takes place through water, much, perhaps, as people detect odors through the air. Pollutants in seawater, even at low concentrations, may disrupt such communication. Hydrocarbons, being natural constituents of seawater, probably do not have a major damaging effect on such communication except when a major spill causes temporary high concentrations of various oil fractions in the water.[28]

Observational Studies

The occurrence of an accidental oil spill presents the opportunity to observe the actual effects on an area. Moreover, the amount and type of oil spilled is usually known. Before summarizing the conclusions from a few of these studies, it would be helpful to establish a set of desired standards for observational studies. These criteria are useful for judging the comparative comprehensiveness and reliability of each report. The desired study standards are: (1) the study is carried on before, during, and after the spill until effects have become negligible; (2) control stations in similar but uncontaminated areas are used; (3) all the components of the oil and their concentrations are measured over time in all their destinations—air, water, on and in the bottom and shore; (4) concurrently, the effects on all organisms that come in contact with the oil are identified and measured; (5) all effects are measured, from subtle debilitation to death; and (6) the effects from the spilled oil are separated from the effects of other natural and man-caused factors.

It is nearly impossible, obviously, to come even close to meeting all these standards, and none of the studies does. Some meet them more closely than others. And their results are considered the more comprehensive and reliable. Each spill is a unique event, however, in terms of amount of oil, character of affected area, time of year, and other features, so it is difficult to generalize on conclusions drawn from different spills occurring under different conditions. Nevertheless, these brief summaries give an impression of what can be affected, by how much, when an oil spill hits an area.

West Falmouth. In September of 1969 the barge Florida went aground near the mouth of a small harbor on the West shore of Cape Cod, Massachusetts.[29] About 650 tons—4500 barrels— of #2 fuel oil, with forty-one percent aromatics, were spilled in a short period of time. As it was during a gale, the fuel oil was rapidly mixed into the shallow water, which ranged from salt marsh to a depth of about forty feet. The oil was also driven up to ten meters into the bottom sediments. The intensely agitated water was filled with clay and organic particles which carried oil to the bottom.

The aftermath of the spill is reported as three phases. First, there was a massive kill of fish, shellfish, crabs, worms, and other crustaceans and invertebrates. The kill was total in areas heavily inundated by the oil. In the second phase, the oil in the sediments spread, through shifting and erosion and being released and returned, to previously uncontaminated areas. The spreading brought damage to sensitive benthic fauna. In the final phase there was gradual repopulation and recovery of bottom life extending inward from less contaminated areas.

The study is primarily of oil in sediments and effects on subtidal benthic fauna. The use of sensitive chemical analysis—gas chromatography—has detected all components of the oil, including low boiling point fractions, even to very low concentrations. The analysis shows that concentrations of most components have persisted in sediments and in benthic fauna up to three years after the spill. The area's 1970 shellfish crop was more contaminated than the 1969 crop. There was no repopulation of bottom living fish and lobsters in heavily affected areas nine months later. Parts of the affected area continue to be closed to shellfishing three and a half years after the spill. The West Falmouth area is not one of chronic oil pollution. So biodegradation of the oil has been slow, due both to absence of bacteria and the burying of the oil.

This is a comparatively reliable study because good prior data on the area were available—scientists were on the scene immediately to document the initial damage, a wide range of sensitive chemical measurements have been made, and the follow-up has extended over more than three years. However, it is limited to sediments and benthic subtidal fauna; observations do not seem to have been made of plankton and nekton or of the more sturdy intertidal life.

The West Falmouth spill study gives a detailed report on some of the effects of a small spill of highly toxic oil on a small area. The spill was characterized by a number of the most damaging possible conditions: (1) shallow water; (2) immediate intense mixing of oil in the water; (3) adsorbent particles and bottom (4) retentive salt marsh; (5) an area highly populated with sensitive organisms, many of which are not mobile or unable to move far; and (6) oil with a high percentage of aromatics, including the water-soluble light and medium molecular weight ones. On the other hand, the area affected is small, facilitating repopulation from surrounding areas. And the time of year was autumn, a time of more adult marine life. Regardless, on a scale ranging from practically harmless to conducive to severe oil spill damage to marine life, the nearshore West Falmouth spill conditions are far toward the worst possible.

Santa Barbara.   Some 80,000 barrels of crude oil escaped through fractures in the Santa Barbara Channel floor for ten days after a well blowout in February, 1969, and in subsequent seepage through the year after the hole was cemented.[30] The spill, originating about five miles off shore, spread rapidly over the water, parts of it blowing ashore in about five days. There was some physical smothering of intertidal organisms, and breeding was reduced in some species. Plankton, including fish eggs and larvae, suffered few effects. The impact on fishing was minimal except to close off areas where there was oil on the water. Channel surveys conducted by state biologists could detect no ill effects on the pelagic fishes in the spill area. These areas had been surveyed in previous work.

Whales and seals appeared unharmed; sea lions were killed, although the net effect on them was difficult to determine because of a high natural mortality of sea lion young. About ninety percent of pelagic birds that came in contact with the oil died.

Little evidence of subtle damage was found except for the five-month departure of small shrimp from the canopies of kelp beds. On the contrary, favorable trends in natural conditions brought improved health in many of the Channel's ecosystems. Recolonization of the intertidal zone commenced after seven weeks, and recovery was nearly complete ten months later.

The Santa Barbara field studies dealt with the planktonic, pelagic (nekton), and intertidal organisms, but not with the subtidal organisms which were the focus of the West Falmouth study. The bottom of the Channel was subsequently covered in places with oiled sediments, but the effects of this on benthic fauna seem to have been minimal. There was prior information on the area, and a variety of scientists were involved in the appraisals of damage. The general conclusion seems to be that damage was physical, not toxic, and that it was localized to heavily coated areas and was temporary.

A number of factors made it difficult to identify the net effects of the spilled oil, an important one being that the Channel is an area of large natural oil seepage and life there may already be well adapted to the presence of chronic oil. Another is that the spill occurred just after a period of unusual storms. There was an extraordinary amount of fresh water runoff into the Channel, changing salinity conditions and carrying with it pesticides, debris, and other pollutants. The stresses caused by the changed salinity may have accounted for much of the damage that was observed. Consequently, the Santa Barbara spill effects seem unique and of limited applicability to the New England area.

Torrey Canyon Spill. On March 18, 1967, the tanker Torrey Canyon went aground on a reef fifteen miles off the southwest coast of England, eventually spilling 700,000 barrels of crude oil, in three discharges, into the English Channel.[31] The effects both at sea and on shore were magnified by the prolific and sloppy use of toxic detergents on the oil. The main goal of the public authorities was to preserve the amenity value of the beaches for the tourist trade. This was generally successful, to the detriment of marine life.

Some observations of the effects of oil unaccompanied by detergents were made. Planktonic animals and fish beneath thin layers of oil appeared healthy. Fishing in the areas affected by the spill did not seem to be noticeably damaged. On rocky shores thick oil smothered some organisms and algae. Coverage with less than one centimeter of oil had little effect on intertidal organisms. Deleterious effects were physical, not chemical, since the slicks were at least eight days old, weathered of toxic components, before they reached shore.

On sandy shores the oil was often buried by seasonal sand accumulations. It did not penetrate far by itself, but did under the influence of detergents. Oil on the sand was eventually washed out by wave and sand motion and was degraded by bacteria. In salt marshes the oil left a black rim. Reproduction of some salt marsh plants was hindered, but, in general, the plants grew through the oil and survived. Recovery of shore areas was rapid where dispersants were not used.

The field studies were carried out primarily on intertidal areas with some observations of plankton and fish made from boats. By and large, sensitive biological and chemical measurements were not taken. But many observations were made by scientists and analyzed against biological data from before the spill. The conclusion appears to be that thick weathered oil has limited physical effects on intertidal life and that other effects are minimal and temporary.

The Brittany Coast of France was visited by a month-old emulsion. It blackened parts of shore areas but had little biological effect.

Other Observational Studies. In 1957 the tanker Tampico Maru dumped 60,000 barrels of diesel oil into a small cove on Baja California when it went aground at the cove's mouth.[32] As with the West Falmouth spill, an immediate heavy kill of numerous species in the cove occurred. Most of the subtidal abalones and sea urchins were lost. A majority of the animals returned by the end of a year, and the area was substantially back to normal after two years.

The collision of two tankers near the Golden Gate Bridge at the mouth of San Francisco Bay spilled 20,000 barrels of asphalt-like Bunker C fuel oil January 19, 1971. The intertidal areas of beaches affected showed heavy kills from smothering. The kills were highest among permanently attached organisms such as barnacles and limpets. Fewer mobile animals were killed, such as crabs and snails, and mussels were not significantly harmed. Most of the heavy oil was gone from the beaches within a year, and recovery was general except for barnacles in large-kill areas where recruitment was slow. In the following summer there was a bloom of brown and green algae in intertidal areas with reduced populations of algae-grazing snails. The study was primarily one of the intertidal area using biologic field observations.[33]

There are other studies of spills which are not summarized here. They do not seem to contradict the impressions that a fresh spill in shallow water that is well mixed can cause immediate kills, that spills allowed to weather cause only limited physical damage by coating, and that recovery to conditions prevalent prior to the spill does occur within a matter of months. These observations are seconded by two scientists who have been monitoring the impacts of Gulf of Mexico offshore oil operations for many years.

Offshore Louisiana Oil Operations. Mackin reports that the effects of both chronic oil discharges and large spills are minimal.[34] But the spills and discharges he has experienced have not been of the one-time massive inundation type. These are not typical of offshore oil operations because blowing wells discharge a ribbon slick over a long period of time. Mackin finds that discharges of wastewater with 30 ppm. oil have a damaging effect only up to 150 feet away from the discharge pipe, the major effects being increased mortality near the pipe and inhibited shell growth of oysters further away. He has also observed benthic fauna before and after spills. His conclusion is that, " . . . crude oil liberated on water surfaces, or incorporated into mud bottoms, lose the toxic elements so rapidly that no effect can be demonstrated. The worst effect encountered was in making oysters taste oily. They retained that taste for short periods only."[35]

St. Amant, who has been monitoring the long run impacts of offshore Louisiana oil operations on the marine environment, has reached similar conclusions:

1) Very rarely will accidental oil pollution have a gross permanent and lasting effect on the ecosystem. Whether minor or long range accumu-

lative effects occur are not readily demonstrable. 2) Seldom is there significant fish or animal mortalities associated with oil spills. 3) Even should extensive and catastrophic loss of animal and plant life occur, once the environment recovers, populations of living things will also return to normal in a reasonable length of time and in some instances in a very short time. This has been frequently observed when serious mortalities occur from natural causes or toxic pollutants. . . . 6) The principal problems associated with accidental oil spills involve contamination of filter-feeding animals (oysters) and esthetic considerations including the fouling of beaches or private property.[36]

The most serious damages caused by oil operations off Louisiana, according to St. Amant, have been the result of the dredging, land filling, spoils depositing, leveeing, and damming that has accompanied oil production in the marshes themselves or has occurred in laying pipelines. These have caused changes in drainage patterns, erosion, salinity levels, and siltation rates. The oil itself, spread out as it is when spilled and eventually weathered and degraded, seems to have comparatively few noticeable widespread adverse impacts on the Gulf area.

Birds

The most obvious direct and immediate casualties of oil spills, expecially of those near shore, are birds. Oil floating on water is a new hazard for them. Like marine organisms, possible damage varies with the habits and the sensitivity of the species. Clark has summarized oil's effects on different types of birds.[37]

The most widespread effect is coating of the bird's plumage by oil, thus causing loss of buoyancy and insulation. In this case the bird may drown or freeze; in the latter case he becomes emaciated from speeding his metabolic rate to keep warm. Birds also swallow the oil in preening. A colony of larger birds, having lower reproductive rates than smaller ones, if it is not wiped out, may take decades to recover. Eggs coated with oil may not hatch. Thus, like marine life, the more prolific are less likely to suffer damage by an oil spill. Since oil, unlike some pesticides and other chemicals, is not persistent in the water, its effect on bird populations is local in nature.

Seabirds, such as auks, diving petrels, and marine diving ducks, which spend most of their time on the water and come to land infrequently, usually for breeding, have the greatest risk. As divers, hunting food under the water, they spend much time socializing on the water. Some are weak fliers, and they dive when disturbed. These species are most likely to come in contact with a slick and to be heavily coated by it.

Other diving birds, such as cormorants, shags, divers, loons, and sawbills, that frequent coasts and coastal waters, are not socializers on the water. Accordingly, there are fewer casualties among them than the seabirds. Sea gulls do not dive. Inhabiting coastal waters, their reaction to disturbance is to fly off. They are strong fliers, so they are not usually lethally affected if brought in contact with a spill. Fish-eating birds that dive from a height, such as gannets and many terns, get heavily oiled if they dive into a slick. However, they generally avoid oiled waters so the risk is lower for them. Waders, which feed on the inter-tidal area of shore at low tide, have been coated by stranded oil on the beach; the damage is not generally serious.

Restoring birds to good health can be a laborious process if they have been heavily coated with oil. It is also frequently unsuccessful. Intensive care is required over a number of months.

Conclusions

On the basis of the studies cited,[38] despite their specific weaknesses, it appears valid to conclude that no significant long-run damage to marine eco-systems comes from oil spills that are infrequent or that have low rates of flow. Recovery has occurred after even the most toxic and concentrated spills, the West Falmouth and Tampico Maru disasters. Massive kills are possible when fresh oil is immediately mixed in shallow water and carried to adsorbent bottom sediments. Spills occurring on open water and spreading rapidly soon lose most of their toxic fractions. Concentrations of these directly under the spill can be high for a few hours or days, but mixing eventually dilutes such components by factors of millions and billions. Thus, although plankton and fish and benthic organisms are killed or sublethally damaged near the spill soon after spillage, the effects do not extend far from where the initial slick started.

Many marine species have extended breeding seasons, produce several broods a year, and, especially in near shore and in shallow water, produce young on a profligate scale to insure survival. Planktonic offspring are distributed over wide areas. Marine life in nutrient-laden, highly oxygenated shallow water, such as that on Georges Bank and near the New England coast, is present in teeming numbers. Thus, an oil spill can be a local short-term disaster, primarily to non-mobile benthic subtidal and intertidal species, but it is not a long-run feature of the environment. Since hydrocarbons, even the complex high molecular weight aromatics, are eventually oxidized and degraded to simpler compounds, to carbon dioxide and water and biomass, the likelihood of oil discharges causing serious long-term accumulating biologic effects appear to be remote.

EFFECTS OF OIL IN WATER ON HUMAN HEALTH

The possible association of oil pollution with human cancer has been raised by Blumer.[39] His chain of reasoning, somewhat simplified, is as follows. A study of oil refinery workers revealed a higher incidence of skin cancer in them than in the general population. Research on cigarette smoking and other possible causes of cancer has shown that some of the high boiling point aromatic hydrocarbons are carcinogenic in animals, being associated with higher than normal incidences of lung cancer and other cancers. Small amounts of such compounds, such as 3,4 benzopyrene, have been isolated in some crude oils. Amounts measured were 450-800 milligrams per ton, about 0.5 to 2 ppm. Such compounds are comparatively persistent in water, neither dissolving nor evaporating, and take months or years to be degraded. They might, according to Blumer, affect man through two mechanisms.

The first is through the food chain. Work after the West Falmouth oil spill has indicated that marine organisms may ingest oil, including these aromatic compounds, and retain it in fat tissue where it is protected from normal degradation. When animals which have ingested and stored such oil are eaten by higher forms of life, these too may become contaminated. Blumer states that, "This has severe implications for commercial fisheries and for human health. It suggests that marketing and eating of oil contaminated fish and shellfish at the very least increases the body burden of carcinogenic chemicals and may constitute a public health hazard."[40]

Several factors, however, indicate that this association of oil pollution in water with increasing the body burden of carcinogenic chemicals is still open to question. First, the MIT Offshore Oil Task Group reports, after a survey of the literature, that, "Incorporation of hydrocarbons themselves apparently does not affect the organisms directly."[41] That is, components of crude oil ingested by marine life do not seem to be carcinogenic to them. Second, there are indications that hydrocarbons in the tissues of marine organisms are in equilibrium with the hydrocarbon content of the water in which they live and are not magnified up the food chain to be possibly concentrated in forms of marine life eaten by man.[42] Third, the amounts of these potentially carcinogenic compounds in oil are, at most, two parts per million. When further diluted in water or in tissues, the concentration may quickly diminish to a few parts per billion. (Heavy metals, pesticides, and other toxic substances are biologically magnified. These are persistent in time and space far longer than hydrocarbons.) Fourth, such minute amounts would appear to have negligible effects because high boiling point aromatics are produced in nature in large quantities. ZoBell reports that polynuclear aromatic hydrocarbons are widely distributed in the sea and are found in fish, seaweed, and widespread marine mud samples. The principal sources of the aromatics are biosysnthesis by anaerobic bacteria, algae, and higher plants, and they are discharged to the sea from land by way of aerial transport and terrestrial drainage.[43] Suess reports on studies showing that phytoplankton produce three tons of aliphatic and aromatic hydrocarbons per square kilometer of ocean per year.[44] Consequently, these compounds are ubiquitous in coastal waters and to lesser extent in oceanic waters, and they are accumulated by some organisms, accounting for significant quantities of 3,4 benzopyrene in fish and shellfish taken from polluted and unpolluted water. These natural quantities of high boiling point aromatics do not appear to have been linked as yet with carcinogenic effects.

Fifth, these hydrocarbons are degraded by some species of bacteria.

The process is slow and requires light intensity to reduce the stability of the compounds, but degradation to simpler compounds does occur.[45]

Sixth, humans are protected from ingesting small amounts of oil-tainted food because the taste of oil is repugnant at a low concentration. The foul tasting components are primarily those containing sulfur. Public health officials generally use the taste test for placing shellfish areas out of bounds or for taking fish off the market both because it protects the public from high concentrations and because the test is simple to administer.[46]

Blumer does not provide detailed analysis of the physiological mechanisms by which these minute amounts of aromatics in marine life might cause cancer in man, nor does he associate concentrations with possible effects. So far the only part of the chain linking oil to cancer in man that has been researched is the presence of aromatics in oil and in minute amounts in oil tainted shellfish. In light of the apparent absence of debilitating effects on the marine life itself, of the doubts about magnification of such compounds up the food chain, of the widespread natural occurrence of these aromatics, of their natural degradation, and of the protection afforded fish eaters by the repugnant taste of oil, Blumer's conjecture appears premature and based on scanty, incomplete evidence.

The other mechanism by which Blumer hypothesizes that oil pollution could cause cancer in man is through skin contact with tar balls and oil in other forms in water. Horn has reported that weathered tar balls can still retain nearly the full range of hydrocarbons of the original crude oil.[47] Blumer states that, " . . . such lumps still contain some of the immediately toxic lower boiling hydrocarbons. In addition, the oil lumps contain all of the potentially carcinogenic material in the 300-500° boiling fraction. The presence of oil lumps ('tar') or finely dispersed oil on recreational beaches may well constitute a severe public health hazard, through continued skin contact. . . ."[48] Again, the only apparent research done on this association has been to document the presence of minute amounts of such compounds in tar balls and in water. The connection is not traced through a chain of dosages and causal mechanisms. Opportunities to research effects of skin contact with oil are plentiful. Natural oil seeps in the Santa Barbara Channel afford the opportunity to study incidence of skin cancer in residents of the area. Gas station workers, auto mechanics, and petroleum production workers come in contact with dosages of oil far higher than those to which any swimmer might be exposed in ocean water. Gross indications of more frequent cases of skin cancer among these populations seem to be lacking. Therefore, at present, while even the causes of cancer are little understood, Blumer's association of oil pollution in water with cancer in man can be regarded as unproved speculation. If the possibility of such carcinogenic effects does exist, it would appear safe to assume that the probability of such effects is remote and of such effects being large and widespread even more remote.

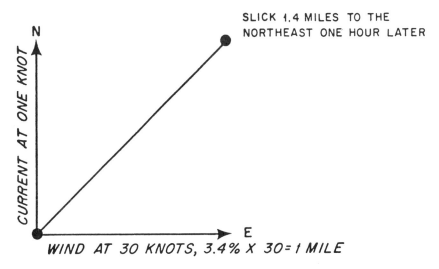

SLICK 1.4 MILES TO THE
NORTHEAST ONE HOUR LATER

N

CURRENT AT ONE KNOT

E

WIND AT 30 KNOTS, 3.4% X 30 = 1 MILE

Figure 4-8 Hypothetical Slick Movement.

POSSIBLE SURFACE MOVEMENT OF OIL SLICKS ORIGINATING FROM GEORGES BANK PETROLEUM DEVELOPMENT

This section examines where a surface oil slick resulting from Georges Bank oil operations might travel. A slick might originate from production operations, in which case it would start from somewhere in the potential oil and gas area on the outer half of the Bank. It might originate from transport operations, from a pipeline to southeastern New England or from a tanker moving to a part anywhere between Delaware Bay and Nova Scotia. Of special concern, because of the amount of damage possible, is the probability of a spill reaching shore. For an assessment of this probability, the dynamics of oil slick movement, currents, and winds must be analyzed.

Dynamics of Oil Slick Movement

As mentioned above most of the low boiling point fractions and the water-soluble compounds of an oil spill have evaporated or dissolved after one to three days. The slick residue becomes more viscous, containing medium and high molecular weight compounds. The wind moves the oil patches, now more cohesive, in addition to dispersing them. Observations of slicks, experiments in wind tunnels, and calculations have produced general agreement that a surface oil slick moves at about three to four percent of the speed of the wind in the direction of the wind.[49]

Wind is the prime determinant of surface water movement, but if the

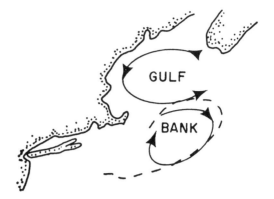

Figure 4-9   General Circulation Directions.

water under a slick is moving at a direction other than that of the wind, the slick also moves at the speed of the water in its direction. In this case, resultant movement of the slick is the resolution of the water and wind movement vectors. Given a northerly current of one knot and a westerly wind of thirty knots, the resultant movement of the slick's center of mass would be 1.4 miles to the northeast, as portrayed in Figure 4-8.

Currents, surface water drift, and winds are highly uncertain natural phenomena; their speed and direction cannot be predicted with accuracy. But general features of wind and water movements for a specific area are known, and historical data are available. Consequently, with the use of statistical techniques, the probability that spills originating on different parts of Georges Bank at different times of the year will stay on the Bank or reach shore can be calculated.

Water Movements

Little is straightforward about water movement on Georges Bank or its vicinity. There is no steady, strong, ever-present current such as the Gulf Stream. The usual water movement over that section of the continental shelf is a drift, influenced by the shape of the bottom, the prevailing winds, and freshwater runoff from land. The drift changes from season to season and from year to year. Some general statements can be made, although their applicability at any time and place cannot be guaranteed.

In general, coastal waters on the continental shelf are in slow circulation at rates ranging for the most part from one to six miles per day.[50] The course of the horizontal circulation is determined largely by the configuration of the bottom, the water rotating in a contraclockwise direction around basins and in a clockwise direction around islands and shoals.

Thus, any strong circulation around the Gulf of Maine is counterclockwise and that on Georges Bank clockwise, as shown in Figure 4-9.

These large eddies are not permanent, however. The Gulf of Maine

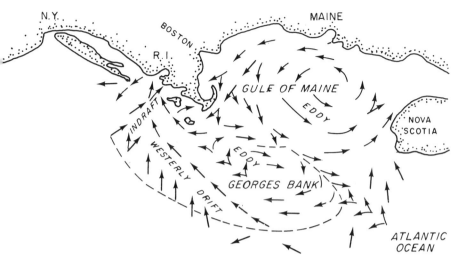

Figure 4-10 General Late Spring and Early Summer Water Movements off the Northeast Coast of the United States. Source: Bumpus and Lauzier, *Circulation on the Shelf,* 1965.

eddy develops rapidly during the spring months. A large cyclonic gyre encompasses the entire Gulf of Maine by the end of May. It slows down in June and by autumn breaks down into a drift southward across Georges Bank. During the spring months an anticyclonic eddy develops on Georges Bank, with its northerly side common with the southern side of the Gulf of Maine eddy. At this time there is a steady westerly drift along the southern side of Georges Bank which travels across Great South Channel. During spring and summer this drift joins an indraft drift toward Narragansett Bay, which is also present seasonally in late spring and early summer. These movements are presented in Figure 4-10.

Consequently, when the Georges Bank eddy and the indraft toward Narragansett Bay are present, the water over the southwesterly part of the potential oil and gas production area moves westerly across Great South Channel and then northward by Nantucket Shoals toward the shoreline between Martha's Vineyard and western Rhode Island. The distance, by the water's route, is more than 100 miles. Therefore, at a net rate of drift of one or two miles a day most of the time, but ranging up to five miles a day, the water takes a number of weeks to reach shore from Georges Bank.

The water over Georges Bank does not generally reach other parts of Cape Cod because one fork of the Gulf of Maine eddy travels southward along the east side of Cape Cod and then westward south of Nantucket Island. There it curls toward the south shore of the Cape or travels on south of Martha's Vineyard.

In late summer the eddy over Georges Bank breaks down. The west side breaks down into a westerly and southerly drift, and the southern part of the

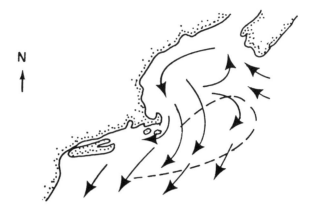

Figure 4-11   General Water Movements off the Coast of the Northeast United States During Late Summer, Autumn, Winter, and Early Spring.

Gulf of Maine eddy breaks down into a southerly drift across Georges Bank. Therefore, during about two-thirds of the year the water from any point of oil production on the Bank is carried offshore and off the continental shelf. Figure 4-11 presents these general water movements during the late summer, autumn, winter, and early spring.

Winds

The winds over Georges Bank and the area south of New England come predominately from the west. Like water movements, they exhibit seasonal variations. The average persistence of the wind from any one direction is only three to six hours. Consequently, a wind rose, which can aid in the calculation of probable wind direction at any point in time, is of limited value in studying the behavior of the wind for periods of weeks. The wind rose can, however, give an impression of general wind directions by season. Figure 4-12 presents monthly wind roses for the Georges Bank.

Autumn and winter winds come strongly and predominately from the northwest and west. They shift southerly during spring and summer but still come predominately from the west. Winds in the cold months are stronger than those in the warm months when calms and light variable winds are more frequent.

The winds that could carry an oil slick from Georges Bank toward shore come from the southeast or are combinations of easterly and southerly winds Over Georges Bank, in the warm months, winds originate from the southeast or east less than fifteen percent of the time, and their average velocity is lower than that from the westerly directions. Therefore, a slick on the water for a month in the warm period of the year would, on the average, be moved to the northeast by the prevailing winds. In the winter months the strong northwesterlies would move

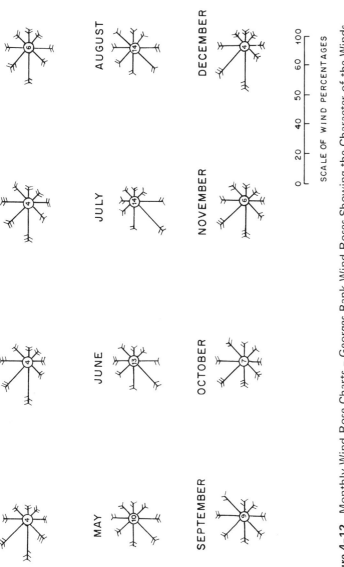

Figure 4-12 Monthly Wind Rose Charts – Georges Bank Wind Roses Showing the Character of the Winds that have Prevailed Over Georges Bank. Source: U.S. Department of Commerce, *Climatological and Oceanographic Atlas for Mariners, Vol. I, North Atlantic Ocean* (Washington: Department of Commerce, August 1959).

Explanation of Wind Roses: The wind rose shows the character of the winds that have prevailed over Georges Bank. Wind percentages are concentrated on eight points. The arrows fly with the wind. Length of the arrow on the scale at right gives the percentage of time the wind blows from that direction. The number of feathers gives average wind force on Beaufort scale. Number in center is % calms and light variable winds.

a slick to the southeast. The influence of the winds on slick movement, then, would appear to move one northeasterly in warm months and southeasterly in cold months. Both directions are away from southeastern New England.

Predicting Possible Paths of Spills Originating from Georges Bank Oil Development

The MIT Offshore Oil Task Group Predictions.   The probability of oil spills reaching shore from Georges Bank in different seasons of the year has been estimated by the Massachusetts Institute of Technology Offshore Oil Task Group.[51] For each of the four seasons the wind is represented as a nine state first-order Markov process with a three hour transition period. The nine states are the eight directions of the compass plus calm. The currents are treated as non-random quantities which depend on location. Four different possible current patterns for the area are used. Then the wind Markov transition matrices and each of the current patterns are combined in the vector equation which gives resultant slick movement. This equation states that the vector of slick movement is given by resolving the vector of water movement and three percent of the vector of wind movement (above, Figure 4-8). The MIT analysts then employed a computer program to simulate the tracks of spills hypothetically released at four different points on Georges Bank. Simulations were made for 200 oil spills originating from each of four possible points with four different current patterns and with Monte Carlo simulations of wind movements for four seasons, a formidable number of calculations. The results seem reasonable, given the current and wind information presented in the preceding two sections, and the conclusions have been checked against drift bottle data by the MIT researchers.

The four hypothetical release points are presented in Figure 4-13. The northwest release point is unlikely to be in the potential oil and gas production area; the remaining three are considered representative of possible spill originating areas. The computer program results indicate that in autumn and winter all spills from the three points would be carried to the southwest or south or southeast, away from shore and off the Bank. Average time of a spill on the Bank would be from three to twenty-five days, the further the originating point to the north and east the longer the stay on the Bank.

In the MIT analysis, the probability of spills reaching shore is calculated for four seasons. Actually, currents flow toward the shore only in late spring and early summer; thus the MIT study overestimates the amount of time the shore is vulnerable. Even with the whole of both seasons considered as a period in which a spill might move toward shore, few such spills are predicted.

For spring, one of the 200 simulations from the southwest release point, using the most likely pattern, reached the Cape Cod-Rhode Island shore. This was after fifty-four days. One of the 200 from the northeast point reached

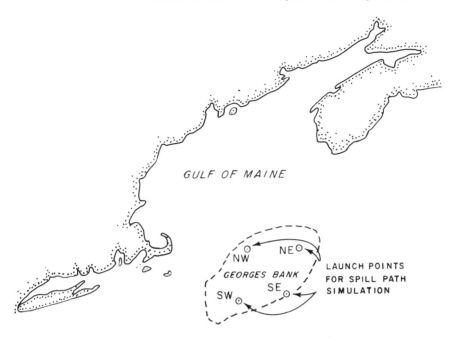

Figure 4-13   Launch Points for Oil Spill Path Simulation in MIT
Study. Source: MIT Offshore Group, *Georges Bank Study,* 1973,
Vol. II, p. 79.

the Bay of Fundy shores after 132 days. All the other simulated paths from the
three points traveled southward out to sea.

   The summer season calculations for this same current pattern showed
ten out of 200 simulated paths from the southwest point reaching the southern
Cape Cod-Rhode Island shore. The quickest spill to reach the shore area took
forty-one days. The average time to reach shore, for the ten, was fifty-seven days.
No other simulated spills from the three points reached any shore during summer;
all were carried southward to the open ocean. Spill paths were tracked for 150
days.

Conclusions
   The MIT analysts report considerable uncertainty about the spill
paths possibly impacting the Bay of Fundy shore areas. Changes in some of the
current assumptions could increase up to ten percent of simulations the number
of spills reaching this area in spring and summer. However, the one simulation
reaching Canada took 140 days. After six weeks a spill is most likely to consist of
a region of small tar balls and scattered remnant patches. After three months in
eastern Gulf of Maine waters it is highly unlikely that the remnants of a spill
from Georges Bank would be noticeable. Consequently, it appears that the shore

Table 4-7. Number of Major Oil Spills That Would Each Have a Five Percent Chance of Reaching Shore over a Forty Year Period (one-sixth of estimates in Table 4-2)

Spill Size Category	Production Estimate		
	low	medium expected number	high upper estimate
Small (less 10,000 bbl.)	0	1	12
Medium (10,000- 100,000 bbl.)	0	1	5
Large (100,000 bbl. plus)	0	1	2

Note: Spill numbers rounded up to nearest whole number.

area that might be affected by an aged spill would be the western and southern Cape Cod-Rhode Island shore. Spills reaching this area could originate on the southwestern part of Georges Bank. That a spill from another oil production area might do so seems a remote possibility. The probability of any single spill from southwestern Georges Bank reaching the southeastern New England shore might be about .005 in spring and .05 in summer.[52] Since the late spring and early summer periods are more likely to see spills reaching shore than early spring and late summer, it can be concluded that there is probably less than a five percent chance that an oil spill originating on the southwestern part of Georges Bank in late spring or early summer could reach the southern Cape Cod-Rhode Island shore after six weeks at sea. Spills in other seasons and from other parts of the potential oil and gas production area would have a negligible probability of reaching shore.

Possible Damage from Georges Bank Oil Spills Reaching Shore

The conclusions reached in the previous section indicate that perhaps five percent of oil spills originating in the southwest part of Georges Bank in late spring and early summer might reach the southeastern New England shore after at least six weeks. With a few conservative assumptions, that is, assumptions tending to overestimate the number of spills reaching shore, the possible number of major oil spills that might travel to Rhode Island or western and southern Cape Cod can be calculated. First, it is assumed that half of all oil production takes place on the southwest part of the Bank. Therefore, half of the estimates of small, medium, and large oil spills presented in Table 4-5 might originate there.

Second, it is assumed that one-third of the spills occur during late spring and early summer (above, Figure 4-3). Accordingly, one-sixth of the number of spills presented in Table 4-5 could occur when there would be a five percent chance of each spill reaching shore. The resulting number of spills is presented in Table 4-7.

Each of the spills in Table 4-7 has a five percent probability of reaching shore. The spills may be treated as Bernoulli Trials: either they reach shore (a success) or not, they exhibit a constant probability of success (0.05), they are independent of each other, and each category of spills has a fixed number of repeated trials. The binomial probability distribution can then be used to calculate the probability, in each category, of no spills, one spill, or more than one spill reaching shore. These are displayed in Table 4-8.

It must be emphasized that the probabilities cited in Table 4-8 are the result of imposing an artificial statistical structure on predictions of the future. However, the upper estimate encompasses many of the uncertainties because it is based on an extermely high estimate of potential oil production from Georges Bank. In addition, the assumptions underlying the calculations tend to over-estimate the possible number of spills reaching shore. Consequently, the chances of a large spill reaching shore are unlikely to be greater than ten percent, of one or more medium size ones less than twenty-five percent, and of one or more small ones less than fifty percent during the possible forty year production life of Georges Bank petroleum fields. The expected and the more likely prob-

Table 4-8. Probablility of Numbers of Spills Reaching Shore from Georges Bank Oil Production Operations during a Forty Year Production Period

	Small Spills (less 10,000 bbl.)	
Probability of:	*Expected Number*	*Upper Estimate*
none	0.95	0.54
one	0.05	0.34
more than one	0+	0.12
	Medium Spills (10,000-100,000 bbl.)	
none	0.95	0.77
one	0.05	0.20
more than one	0+	0.03
	Large Spills (100,000 bbl. plus)	
none	0.95	0.90
one	0.05	0.10
more than one	0+	0+

Source of Probabilities: Frederick Mosteller, R.E. Rourke, and George B. Thomas, Jr., *Probability with Statistical Applications* (Reading, Mass.: Addison Wesley, 1970), Table 4.

ability is about five percent for one of each size spill to reach shore in forty years.

However, the MIT analysis of simulated spill paths indicates that should one reach shore it would be after at least six weeks and most probably after eight weeks. By that time, assuming no encounters with storms, the remnants of the spill would be, at worst, in the form of widely dispersed tar balls and, perhaps, small patches of oil in water emulsion. It would not be an intact continuous slick.

The environmental impacts of such dispersed and weathered remnants would be much less damaging than those of a fresh oil spill near shore. Since the remnants would be scattered, no single area of the shore would be covered. Small pockets of intertidal life might be coated, and some smothering of non-mobile organisms could occur. But most of the toxic fractions would have evaporated or dissolved weeks before impact on shore. Therefore, the damage to shore forms of marine life would be minimal.

The major impact of the weathered remnants of the spill would be on the sensibilities of those people stepping on and seeing the tar balls and emulsion patches on the beach. In the "worst credible" case the remnants might begin washing up on the western and southern Cape Cod and southern Rhode Island shore just before the Fourth of July weekend after much publicity about the spill's travels at sea. In this case a number of people might forego visits to the affected area, and those who do visit would notice the pollution and experience decreased enjoyment.

The probability of such an occurrence appears to be less than five percent over a forty year period. It would be the maximum possible risk to shore resulting from oil production operations on Georges Bank.

Possible Damage to Georges Bank Marine Resources from Major Oil Spills and Chronic Oil Discharges

The Georges Bank Fishery's Prospects.   If present trends continue, there may well be no commercially exploitable marine resources on Georges Bank within a few years. In that eventuality, major oil spills and and any other petroleum production activities would have a negligible impact on the fishery. In the early 1960s large fleets of foreign fishing ships, including factory ships and giant trawlers, arrived on Georges Bank after depleting the resources of other areas closer to home ports. Prior to their arrival the area had been fished by small boats from New England and Canada. The total catch taken from the Bank increased rapidly in the 1960s, as portrayed in Figure 4-14. The total tons caught mask the decline in the more desirable species which have already been seriously over-fished, especially the haddock. These figures are for Subarea 5 as designated by the International Commission for the Northwest Atlantic Fisheries (ICNAF), the fifteen nation body that was formed to regulate the fisheries. The totals have been made up with the less valuable species which are usually made into fishblocks and fishmeal. The trend

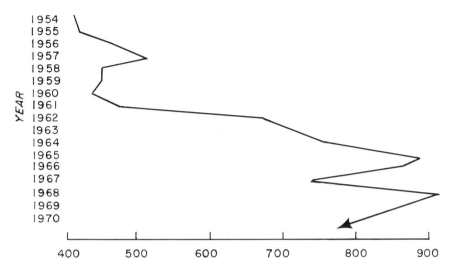

Figure 4-14   Total Catch, ICNAF Subarea 5, 1954–1970 (in 000 metric tons). Source: International Commission on Northwest Atlantic Fisheries, *Statistical Bulletin, Vol. 20, for 1970* (Dartmouth, Nova Scotia, Canada: ICNAF, 1972).

of the haddock catch is portrayed in Figure 4-15. Haddock and possibly other species have probably already been fished below the point from which they might make a recovery on the Bank. The total catch of most species is well above the maximum sustainable yield. The catch of silver hake is on a precipitous decline similar to that of haddock, and cod is on a similar but lagging trend downward. The herring catch has gone from about 70,000 metric tons in 1960 to a peak 407,000 metric tons in 1968 and is now also on a downward trend, with 220,000 tons caught in 1970. The total catch, by species, for ICNAF Division 5Ze, which covers Georges Bank and is the main part of Subarea 5, is presented in Table 4-9 along with the United States share of each catch. The foreign nations involved in exploiting Georges Bank and their shares of the total 1970 catch are presented in Table 4.10. There are numerous problems with these data. East Germany is not a member of ICNAF, for example, and significant underreporting is probable. But these numbers do show which species are commercially important and which nations are involved in overfishing Georges Bank.

There is every indication that the large foreign fleets will continue to fish on Georges Bank until such harvesting is no longer profitable.[53] International management has failed, even with imposition of national quotas on catches of threatened species. Quotas have been set so high that some of them have not been filled. Since the Polish, West German, and Soviet governments are amortizing their large factory ship fleet investments over a short period of time, they have no

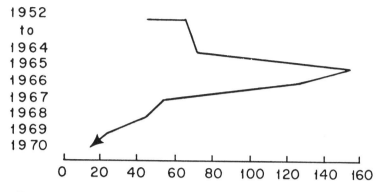

Figure 4-15   Total Haddock Catch, ICNAF Subarea 5, 1952-1970

incentives to restrict fishing on the East Coast of the United States. It appears that only single-nation management of the fishery by the United States and/or Canada could prevent the total depletion of commercially valuable marine resources on Georges Bank within a few years. A 200 mile national jurisdiction out to sea might provide this, which is a possible outcome of the Law of the Sea Conference scheduled for 1974. But the probable setting for future oil and gas development on Georges Bank is in a fishery depleted of commercially desirable species and unable to recover for many years.[54]

Table 4-9.   Nominal Catch in ICNAF Division 5Ze, 1970, by Species, (in metric tons fresh) Total and U.S. Catch

All Species			
Groundfish		*Pelagic Fish*	
silver hake	25,000	herring	170,000
US	4,000	US	300
yellowtail flounder	25,000	mackerel	65,000
US	24,000	US	400
cod	24,000	other fish	19,000
US	13,000	US	1,000
haddock	11,000		
US	8,000		
winter flounder	7,000		
US	7,000		
pollock	4,000	sea scallops	46,000
US	2,000	US	12,000
other groundfish	12,000	other shellfish	4,000
US	6,000	US	1,000

Source: *Ibid.*

Table 4-10. Total Catch by Country in
1970, ICNAF Division 5Ze (metric tons fresh)

Poland	97,000
Soviet Union	94,000
West Germany	83,000
United States	80,000
Canada	42,000
Japan	5,000
Spain	1,000

Source: *Ibid.*

Maximum Credible Damage to Fishery from Oil Spills

An assessment of the maximum credible risk to the Georges Bank fishery from oil operations must assume that the fishery is being exploited at a sustainable rate. The Bank is a highly complex and abundant ecosystem.[55] As on land, its plant life is scanty from late autumn to early spring. Then the vernal augmentation, a flowering of phytoplankton, occurs, climaxing in April. The rich flora persists in enormous quantities on the Bank through June or July and flowerings gradually come to a halt in autumn. The zooplankton populations naturally vary with the phytoplankton, as that is their food supply. The zooplankton are in turn fed upon by some fish of the area.

The two pelagic fish of commercial importance, herring and mackerel, spawn in vast numbers along the shore and off shoals. Most of Georges Bank is shoal, with two large ones, Cultivator Shoal and Georges Shoal, taking up the north central part of the Bank. Herring are weak, small fish which travel in widely dispersed schools, wintering in deep waters. Their eggs, averaging 30,000 per female at a time, sink to the bottom in layers during the August-November spawning period. Mackerel eggs, on the other hand, are buoyant. During spawning period from April through August a female can release 40,000-50,000 eggs per spawn. Mackerel spawn anywhere in their wanderings, schooling in large numbers and carrying out seasonal migrations, into shallow waters in spring and into deep water in autumn. Mackerel are swift swimmers and powerfully muscled. Both herring and mackerel vary widely in local abundance from year to year, one year appearing in vast numbers over Georges Bank or Nantucket Shoals and in other years appearing elsewhere, having disappeared from Georges Bank.

The area's high-value groundfish are also prolific breeders; each female produces hundreds of thousands of buoyant eggs during her lifetime. The cod is especially prolific, releasing millions of eggs a year. The eggs and subsequent larvae float about helplessly for a few months until the young fry seek bottom. Haddock, cod, and flounder spawn from February to April, when the water is

Figure 4-16   Habitats of Major Species on Georges Bank.

coldest, over the shoal areas of the Bank. As the water warms they move to cooler, deeper water, returning to shoal water in autumn.

The groundfish live primarily within two meters of the bottom, traveling in loose schools. Haddock, cod, and flounder are abundant mostly on the eastern half of the Bank. Silver hake range over the entire Bank. A warm water fish, they appear in late spring in shoal water and journey offshore for wintering. Figure 4-16, shows the principal habitats of the major species on Georges Bank. The species mentioned here also spawn and feed over other areas often ranging from south of Cape Cod north to the Grand Banks and across the North Atlantic.

Sea scallops, the predominant shellfish harvested from Georges Bank, inhabit bottoms from thirty to eighty meters deep. They are concentrated in Great South Channel in forty to eighty meters of water and, to a lesser extent, in the deeper water of the southwestern part of the Bank.[56] The recently found northern lobster fishery is concentrated in the deep water on the southern edge of Georges Bank. In late spring the lobsters move toward shoals and coastal waters to spawn over a wide area from Georges Bank to Long Island.[58]

It is clear from this brief discussion of the marine life of Georges Bank

that the reproductive strategies of the species there are to produce young in prolific numbers to insure that some survive. The planktonic larvae are eaten by surface feeders, and many are swept out to sea where they cannot reach bottom or
find food. An oil spill could kill a number of the larvae, but this would be a local
stress both in place and time. The eggs are spawned over a number of months
over a wide area. Therefore, any damage done by a spill or number of spills would
not affect the populations to a noticeable extend since mortality of young is
naturally very high. The MIT Offshore Oil Task Group reports that in their
"worst case" analysis, a spill of about 700,000 barrels of oil (Torrey Canyon
size), less than two percent of the year class of larvae of Georges Bank fish
species would be damaged. They conclude, "In summary, it appears unlikely that
a single spill could kill enough larvae to have a noticeable effect on adult populations. Nature appears to have provided a reproductive process which is relatively
insensitive to very short run phenomena."[58] The discharge of chronic amounts
of wastewater with about 50 parts per million of oil from oil production
platforms could damage larvae in the vicinity of the platform. However, the area
of concentration above 0.1 ppm. would extend at most a few thousand feet from
the platform[59] and the number of oil production platforms is unlikely to exceed
ten. Therefore, the MIT Offshore Oil Task Group concludes:

> The largest such area [with levels of oil above 30 parts per billion near
> a platform] is in the neighborhood of two square miles. Even ten such
> platforms could affect only a very small proportion of the larvae of
> even the more concentrated species. At present, it appears unlikely
> that the continuous discharges of a very large, high water production ·
> discovery would have a noticeable impact on the fishery. This find-
> ing is consistent with Gulf experience.[60]

Adult fish are able to avoid oil in or on water. After sensing low levels of
it they avoid the contaminated area. Furthermore, being adult and strong, they are
not as sensitive to the toxic effects of oil as larvae are. The predominant species on
Georges Bank, the groundfish, would be unlikely to come in contact with high concentrations of oil because they inhabit the water near the bottom where any oil
fractions would be highly diluted. Consequently, it appears that neither large oil
spills nor chronic discharges of low amounts of oil would appreciably damage adult
fish on Georges Bank. In the "worst credible case" light oil might be immediately
driven into shallow water by a storm. Much of the toxic fractions would dissolve
at once, killing or debilitating any fish in the area. But the area would be small—a
couple of square miles at most—and if the water is mixing intensely the oil would
also be diluted quickly. Since Georges Bank is in the open sea, there are no enclosed waters where oil fractions might persist at high concentrations.

The northern lobsters inhabit deep water and spawn over a wide area,
so a temporary surface oil spill would not seriously affect them. Since they live
on the edge of the continental shelf in water more than sixty meters deep, oil operations above their habitats are unlikely in the near future. Increasing levels of oil

in the water might disrupt chemical communication between the lobsters, but concentrations at which this occurs are unknown and the effect itself is still open to question. Elevated levels from a large spill would be temporary. Levels from chronic discharges a few thousand feet from a platform would already be in parts per billion, concentrations typical of ambient levels in the deep ocean which are too low to cause any direct effects.

Sea scallops are the species most sensitive to damage from oil discharges into the water. As filter feeders, they concentrate parts per billion of oil into levels that can cause tainting, making them inedible. Oil operations are not likely over the northern part of Great South Channel, one of their major habitats. But they are possible over their other major habitat area, the southwestern part of the Bank. The "worst credible" damage would result from stationing the ten platforms, all producing large amounts of oiled wastewater, in that area. This would chemically contaminate twenty square miles (ten platforms times two square miles) of the bottom with thirty parts per billion of oil or more. Any sea scallops there might concentrate the oil and, while not being affected themselves, they could become inedible.

Assuming that sea scallops are not soon overharvested below their recovery population,[61] that the southwest Bank area accounts for about half the total catch, and that the sea scallops are distributed there over an area of about 500 square miles,[62] the gross economic value of such damage can be calculated. Hypothesizing a high sustainable total yield of twenty million pounds a year, the twenty contaminated square miles might prevent the harvesting of four percent (20 divided by 500) of the possible catch in the southwest Bank area or two percent (4 percent times one-half) of the total sustainable catch. This would be approximately 400,000 pounds. Valued at the current dockside price of $1.80 a pound, the gross value of the damage to this expensive resource could be, under a series of "worst case" conditions and assumptions, about $720,000 a year.[63]

The effects of a large oil spill driven into the water would be temporary because the scallops can cleanse themselves after the elevated oil concentrations are diluted. Consequently, the greatest major possible damage that might be sustained by the Goerges Bank marine resources would be the contamination of about twenty square miles of sea scallop grounds on the southwestern part of the Bank. Possible damage from both spills and chronic discharges of oil on other species would be negligible. This conclusion also seems reasonable in the light of studies of the impacts of oil operations in the Gulf of Mexico[64] and North Sea[65] fisheries, which indicate no substantial related damage to fishery productivity.

Effects of Publicity of Oil Spills on the Fishing Industry

Clark points out that, "The fishing industry is vulnerable to the vagaries of mass psychology in a way that few others are and even the suspicion that fish may be contaminated is sufficient to depress the market."[66] After the Torrey Canyon spill fish sales in Paris dropped by half, regardless of the source of the fish.[67] Similar occurrences occurred in New England after "red tides." Therefore, the major damage to the fishing industry from a major oil spill on Georges Bank would not come from the actual physical damage done to marine resources. The reporting of the spill, especially if accompanied by pictures of dead fish and birds, would trigger a reluctance on the part of many to eat any form of seafood. This may persist for weeks, temporarily depressing the market for all seafood, especially for fresh fish and shellfish. A subsequent section deals further with the role of the news media in reporting environmental damage.

Wastes. Drilling muds, solid wastes, and sewage can be discharged into the ocean, in addition to oil, during oil and gas production operations. Drilling mud is a mixture of water, clay, and chemicals used to control wells during drilling. The mud is circulated in a closed system, so in normal operations little is lost. Interior Department OCS Orders prohibit lessees from dumping drilling mud containing oil into the sea. Many companies now use water-based drilling mud which can be disposed of at sea provided toxic substances are neutralized prior to disposal. Compliance with the regulations vary, and it can be expected that some mud, either accidentally or intentionally, would be discharged onto Georges Bank during initial well drilling. However, no toxic effects on marine life have been attributed to such muds, and they are highly diluted in the sea. If damage does occur, it would be localized to the vicinity of the drilling operations. Filter feeders in the area might be damaged by the increased turbidity.

Drill cuttings, sand, and other solids may be discharged to the water during drilling and workover operations. OCS Orders require them to be cleaned of oil before dumping. The material itself is natural and would have little adverse impact except for possible physical effects on benthic organisms in the local area. Other solid waste materials, including mud containers, are required by Interior regulations to be incinerated or transported to shore. Even if dumped, such material is by and large chemically inert. However, floating debris can present an unpleasant sight and potential hazard to any boaters in the area.

Finally, the sewage from men working on the platforms would be a small daily amount discharged from scattered locations. It could be regarded as a marine nutrient, as human waste contains many substances utilized by marine life. Nevertheless, OCS Orders require sewage treatment on platforms. It appears that with the stringency of current regulations on lessees, and probably even without them, the impacts of these wastes from oil and gas production operations would be minor, temporary, and limited to the vicinity of the platforms and drilling vessels.

Structures. Oil and gas production on Georges Bank would require a number of fixed structures. The possible number of platforms is uncertain, but twenty would be high and fifty a remote possibility. Fishermen generally stay a quarter to a half mile from them for safety reasons and to avoid damaging their nets. Therefore, these structures would prevent fishing on up to 150 square miles of the Bank (50 platforms times 3 sq. mi. each). Given the overfishing of the area, this could be considered a favorable conservation measure. The impact, either favorable or detrimental, would be small since the Bank is only about 10,000 square miles in area.[68]

Collisions with commercial ships is another possibility. The U.S. Army Corps of Engineers does not permit structures to be built in commercial navigation lanes; therefore, the only danger would be from ships off course. In fifteen years of experience in the Gulf, which has more than 2,000 structures, one ship collision has caused an oil spill. Three or four fishing boat collisions have also been reported, none causing a spill.[69] Therefore, damage in collisions with platforms is slight. The offshore structures, on the other hand, may serve as havens to ride out a storm or to make repairs. In conclusion, it is not apparent that the environmental effects of offshore platforms would be unfavorable. They would be at least seventy-five miles offshore, out of sight of land. Impacts on the fishery could be slightly favorable.

Conclusions. The possible adverse environmental impacts from oil and gas production operations on Georges Bank appear to be small. There seems to be less than a five percent probability that, during a forty year production period, the scattered and weathered remnants of a large spill originating on the Bank might reach shore—and even then it will have been six weeks at sea. The primary stress on the marine resources of the Bank is likely to be systematic overfishing by large Polish, Russian, and West German fleets. In a few years the numbers of several commercially important species may be reduced below possible recovery levels, and haddock apparently already has been so depleted. However, should management of the fishery become successful, the marine ecosystem is so prolific and resilient that the effects of major oil spills, chronic oil discharges, and dumping of other wastes from production operations would be small and limited to the immediate vicinity of the spill or platform. Under a series of "worst case" conditions and assumptions, damage to the annual sea scallop crop from chronic oil discharges could reach two percent of the sustainable yield. The "worst credible" damage to other species would be minimal.

ENVIRONMENTAL IMPACTS OF TRANSPORTATION OF GEORGES BANK OIL AND GAS TO SHORE

Transport Possibilities, Routes, Destinations

Gas would be transported to shore from Georges Bank only by pipeline, since there is no economical method for short-term liquefaction of the gas for tanker transport. A leak from a gas pipeline, while possibly causing safety hazards, would not cause any adverse impacts on marine life since it bubbles off and evaporates in the air.

Oil, however, can cause damage to marine life; it is transported to shore either by pipeline or tanker. As of 1971, only about three percent of Gulf OCS oil production was moved to shore terminals via barge.[70] Pipelines are by far the preferred mode because they are less expensive per barrel moved, are safer, and are more dependable as oil can be moved in any type of weather. Therefore, if rights-of-way for an oil pipeline can be obtained, and if shore terminals are available a reasonable distance from production, Georges Bank oil would be brought to shore via pipeline. Few trunk lines from offshore collection points to land exceed 100 miles in length, but technology may be available in the near future to extend such lines up to 250 miles.[71]

On the Bank, gathering lines would move oil from dispersed fields to a central pumping station. Trunk lines to move the oil to shore have outside diameters of up to forty-eight inches and carry up to 250,000 or even 500,000 barrels a day. Consequently, one such line would probably suffice for all Georges Bank oil production.

Under a number of possibilities, however, tanker transport of Georges Bank oil production might be preferred over a pipeline. If production is small—less than about 10,000 barrels a day—or if it peaks and declines over time, the fixed capacity of a pipeline would not be economical, and tanker transport, which is more flexible, would be chosen. If no oil pipeline right-of-way can be obtained over state-owned submerged lands or if no shore terminal area can be obtained, tanker transport would be a necessity. If no refinery is near the shore terminal, costly coastal transport to a distant refinery area might be required. Accordingly, direct tanker transport from the production area to a refinery complex outside New England might be the less expensive alternative.

All these conditions are possible. Small oil production, lack of refineries in New England, and intense opposition to an oil pipeline by state and local officials and citizens seem highly probable. (This will be discussed in a subsequent section) Therefore, tanker transport of Georges Bank oil, if any is found, to a destination outside New England is as least as likely, and perhaps more likely, than oil pipeline transport to southeastern New England. Possible pipeline destinations and ports for tankers are presented in Figure 4-17. This is a partial list since unforeseen areas could be developed as refinery areas in the future. If the refinery situation on the East Coast remains static, Georges Bank oil might be taken by tanker to refineries in the Bahamas, further into the Caribbean, or even around to the Gulf Coast. The more likely destinations are nearer at hand.

Oil Tanker and Barge Traffic to New England Ports

Transporting Georges Bank oil to shore in New England would not create a new environmental risk of major oil spills. Since no petroleum produc-

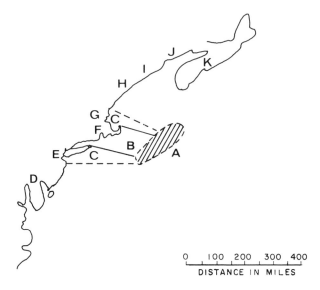

Key to Areas:

A   Potential Georges Bank Oil Production Area
B   Possible Oil Pipeline Routes – 150 mile maximum length
C   Possible Oil Pipeline Routes – 250 mile maximum length[1]
D   Delaware Bay – refinery area, possible site of deepwater
    oil import terminal
E   New York-New Jersey refinery area, port
F   Narragansett Bay – only refinery in New England (very small),
    port for oil products
G   Boston
H   Portland, Maine – large port for crude oil imports for Canada,
    possible refinery site
I   The deepwater bays on the Maine Coast – possible import ports
    and refinery areas
J   St. John Deep, New Brunswick – developing port for supertank-
    ers and refinery area
K   Halifax, Nova Scotia – port and possible refinery area near off-
    shore petroleum development there

[1] North of the C area are waters of basins and the Northeast Channel
more than 220 meters deep, too deep and expensive for a pipeline.

Figure 4-17   Possible Oil Pipeline and Tanker Destinations.

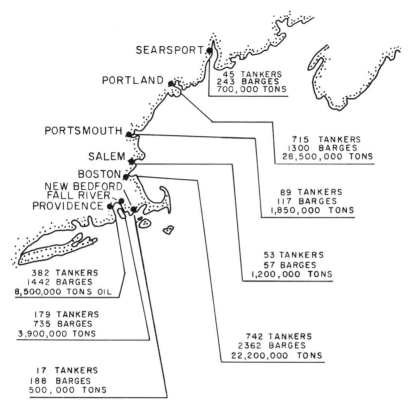

SEARSPORT
PORTLAND
45 TANKERS
243 BARGES
700,000 TONS

PORTSMOUTH

715 TANKERS
1300 BARGES
28,500,000 TONS

SALEM

BOSTON
NEW BEDFORD
FALL RIVER
PROVIDENCE

89 TANKERS
117 BARGES
1,850,000 TONS

53 TANKERS
57 BARGES
1,200,000 TONS

382 TANKERS
1442 BARGES
8,500,000 TONS OIL

179 TANKERS
735 BARGES
3,900,000 TONS

742 TANKERS
2362 BARGES
22,200,000 TONS

17 TANKERS
188 BARGES
500,000 TONS

Notes: Connecticut ports not included.
Includes transshipments.
Ten small traffic ports not included.
Tanker and Barge traffic is full ones both arriving and
departing.

Figure 4–18 1970 Oil Tanker and Barge Traffic at New England
Ports. Source: MIT Offshore Group, *Georges Bank Study,* 1973, Vol.
I, p. 88.

tion now originates in New England, and since no oil pipelines on land extend
there, all of New England's refined oil products, except for 7,500 barrels a day
from a small Mobil refinery in Providence, arrive by barge and tanker at a number of product distribution ports.[72] This barge and tanker traffic, with numbers
of tons of oil carried (about seven barrels per ton) in 1970, is presented in Figure
4-18. About ten small ports are not included. The islands are dependent on barge

traffic for their fuel. Nantucket received twenty-four barges with 27,000 tons of oil and Marthas Vineyard 111 barges with 35,000 tons in 1970.

The large barges come from New York and Delaware Bay, while small ones are used to transship oil from the larger to the smaller New England ports. Roughly half the tanker traffic is from New York and Delaware Bay and half from foreign refineries. Nearly all the barge and tanker traffic heading to ports north of Cape Cod, from Rhode Island, New York, and points south, travel by way of the Cape Cod Canal, Thus, most of the coastal shipping creates spill hazards for southeastern New England.[73]

Net Changes in Risks of Nearshore Oil Spills

From this information it seems clear that at present, and for the near future, the shipment of oil to New England creates the hazard of major spills near and on shore. It is assumed here that this hazard is proportional to total tonnage of oil carried along the coast. If Georges Bank oil production is shipped by tanker to a refinery area outside New England, the transport operations would result in no increased spill hazard to New England shores. If the production is shipped to a new refinery in New England, from which the products would be marketed by land, the Georges Bank production would displace already existing barge or tanker shipments and would also result in no new oil spill hazards.

However, if the oil is carried by pipeline to southeastern New England and then shipped by tanker or barge out of the region for refining, this procedure would result in a net increase in coastal oil shipments and a concomitant increased risk of oil spills from pipeline or ship. The likely amount of Georges Bank oil production would be less than 3,400,000 tons a year, but it could range up to 22,000,000 tons per year (above, Table 2-6, seven barrels per ton). Therefore, the increased risk from sea shipment of oil would probably be somewhat less than that from 1970 oil tanker and barge traffic to Fall River, which was 3,900,000 tons. On the other hand, it could range up to the equivalent of 1970 oil traffic in and out of Boston Harbor, 22,200,000 tons. By the time Georges Bank production could be available—the late 1970s or later—such oil shipments to these ports would have increased, possibly by as much as 100 percent (above, Chapter Three). Total 1970 coastal distribution of oil in New England was about 100 million tons. A doubling by 1985 would result in oil traffic of 200 million tons. The barge and tanker traffic that might take Georges Bank oil out of the region from the pipeline terminal would then increase traffic by probably less than two percent (3.4 divided by 200) but possibly up to eleven percent (22 divided by 200). It must be emphasized that this increased environmental risk would occur only if the oil is carried to the New England shore and then shipped out of the region.

If the pipeline-transported oil is shipped out of the region, a risk of spills from such a pipeline would occur in addition to that from the coastal shipping. Pipeline transport of oil is generally safer than tanker or barge transport for a number of reasons: (1) since the pipeline does not travel in the water, it cannot cause a collision; (2) opportunities for human error are limited; (3) adverse weather does not affect oil transport; (4) oil transfer operations at shore are simpler; (5) pipelines are now required by the Interior Department to be buried in up to 200 feet of water, making anchor dragging accidents less possible; and (6) automatic shut-in of the line is possible once a leak is detected by observation or by automatic pressure-sensing devices which can activate shut-in.[74] Therefore, the risk added by pipeline transport would not be as great as that from the coastal shipping.

An approximate factor to use in comparing pipeline and shipping spills is derived from comparing the number of spills per amount of oil moved by each mode in U.S. waters from 1964 to 1971. Imports of oil from abroad by tanker were about 1.7 million barrels a day in 1964 and 3 million in 1971.[75] United States production during the same period went from about 9 to 11 million barrels a day. Assuming half of this was shipped along the U.S. coast as crude oil or products, then tanker and barge traffic carried about 4.5 million barrels a day in 1964 and 5.5 million in 1971. Imports by sea and U.S. intercoastal traffic then added up to 6.2 million barrels a day in 1964 and 8.5 in 1971. Assuming that all federal OCS oil production was moved to shore by pipeline, transport was about 0.8 million barrels in 1964 and about 2.5 in 1971.[76] This was, respectively, about thirteen percent of all ship traffic in oil in 1964 (0.8 divided by 6.2) and thirty-four percent in 1971 (2.5 divided by 8.5), an average of about twenty-three percent (13 plus 34 divided by 2), or, roughly, one quarter.

There were three major OCS oil pipeline spills in the period 1964-1971 (above, Table 4-1) and about twenty barge and tanker spills in U.S. waters.[77] The largest spill, of more than 200,000 barrels of #4 fuel oil, was caused by hull failure of the Keo 120 miles southeast of Nantucket. The largest OCS pipeline spill was 160,000 barrels. Therefore, the possible range in size of spills can be considered comparable, although imports in supertankers would create the risk of spills as large as two million barrels. The OCS pipeline spills were fifteen percent of the shipping spills (3 divided by 20). Since OCS oil transport by pipeline was about twenty-five percent of that moved by ship, a rough factor for comparing likely numbers of spills would be sixty percent (4 times 15 percent); that is, for the same amount of oil moved, pipelines might experience sixty percent of the number of possible tanker and barge oil spills.

This generally confirms the conclusion that pipelines are safer than ship transport. The sixty percent factor itself is based on limited historical data and could vary considerably for specific locations and time periods. However, it can be used to calculate the total added risk of oil spills if Georges Bank oil is piped to shore and then shipped out of the region. Since the

Table 4-11. Net Change in Risk to New England of Nearshore Oil Spills Resulting from Transport of Georges Bank Oil and Gas

Scenario	Oil transport		Gas transport	
	medium estimate	*high estimate*	*medium estimate*	*high estimate*
Oil pipelined to shore and shipped from New England.	+3%	+18%	−1%	−3%
Oil taken by tanker to New England.	0	0	−1%	−3%
Oil pipelined to refinery in New England.	−0.8%	−4.4%	−1%	−3%

shipping would increase the risk by probably less than two percent but possibly up to eleven percent, the pipeline could be considered to add sixty percent of these percentages for a total added risk from ship and pipeline of probably less than three percent but possibly up to eighteen percent.

Under other conditions, transport of Georges Bank oil would reduce the total risk of nearshore oil spills to New England. If the oil is piped to a refinery in New England it would displace tanker traffic to the area. This would reduce the risk of spills by probably about 0.8 percent (40 percent of the tanker risk of 2 percent) but possibly up to 4.4 percent, (40 percent times 11 percent).

The production of gas on Georges Bank and its transport to New England would probably reduce the risk of nearshore oil spills, because it would be substituted for imports of low-sulfur residual fuel oil.[78] The oil equivalent of Georges Bank gas would be about 1.6 million tons a year and could be up to 6.3 million tons.[79] These are, respectively, about one percent and three percent of the 200 million tons of oil that might be carried by ship to New England in 1985. Therefore, the transport of Georges Bank gas to New England would probably reduce the total risk of nearshore oil spills by one percent, possibly up to three percent.

Conclusions
New England experiences a high risk of nearshore oil spills because all of its oil is carried to distribution ports by tankers and barges. In 1985, about 200 million tons of oil will be shipped along the New England coast in thousands of tanker and barge trips. If movement of tons of oil to shore and in and out of

ports is considered a reasonable proxy for the risk of nearshore oil spills, the net change in total risk to New England from Georges Bank oil and gas transportation can be calculated. These predicted changes are summarized in Table 4-11 for three different transport scenarios.

In the unlikely event that large reserves of oil are found on Georges Bank, and if this oil is pipelined to New England and then shipped out of the region for refining, which is possible, the risk of nearshore oil spills to the New England coast could increase up to eighteen percent. If small finds of oil occur and if small or large gas deposits are discovered, the net effect on oil spill risks would be small—probably a decrease in risk of a few percent. At the other extreme of possibilities, if a large amount of oil and gas is found and if the oil is pipelined to a refinery in New England, the total risk of nearshore oil spills could decrease by up to seven percent.

The possible adverse impacts of nearshore oil spills on the New England coast are not analyzed here because Georges Bank development would most likely have a small net effect on the number and size of such spills. General impressions of potential impacts can be made from the reports on past spills, especially those on the West Falmouth, Torrey Canyon, and Santa Barbara spills.[80]

COPING WITH OIL SPILLS

The preceding analysis of possible oil spill damage has not considered measures to cope with a spill to decrease attendant damage. Several are available: containment and recovery, dispersion, sinking, burning, and other methods. Each measure, or combination of measures, can vary widely in its effectiveness, in cost, and in possible damage to marine life caused by its use.[81]

The National Petroleum Council, an oil-industry advisory group to the Interior Department, is pessimistic about the capability of containment devices or absorbents to contain a spill at sea. Their report states, "Containment devices that will restrict the movement of oil in the open sea are not available. There have been no demonstrations of oil-recovery devices with the ability to pick up oil from large spills in rough waters at the needed rates and efficiencies."[82] On adsorbents they state, "Generally speaking, sorption with either natural or synthetic materials presents logistical problems of both dispersal and collection which are not adequately met with present equipment."[83] These methods—containing a spill with booms or the broadcasting of adsorbents such as hay or polyurethane foam chips and recovering the oil with skimmers or rakes—are sometimes successful in protected waters. But the current state of the art cannot handle waves higher than five feet, and the equipment generally cannot be deployed far offshore. Interior Department data indicate that recovery of oil spilled on the OCS has ranged from about one percent in the 1969 Santa Barbara spill to forty percent for the late-1970 Shell spill. These were, however, within ten miles of shore, much closer than Georges Bank operations would be. Containment and recovery equipment might be deployed with success in the event of nearshore spills occurring in calm weather.[84]

Dispersants are chemicals which, much like laundry detergent, act to disperse oil into the water in fine droplets that do not coalesce again or rise to the surface. Dispersed oil does not "wet" surfaces, nor is it visibly noticeable. Therefore, although dispersants can make oil more available for ingesting by marine life and may, in shallow and restricted waters, cause significant damage to marine life themselves, their use can prevent an intact oil spill from reaching and coating shore. If such damage to shore scenery and property is considered more critical than the possible increase in damage to marine life, then with sufficient mixing by spraying and boat propellers, the use of dispersants can serve to place much of an oil spill out of sight. Sinking agents may serve a similar purpose by adsorbing floating oil and carrying it to the bottom. This, of course, may damage benthic organisms, especially filter feeders, but sufficient spreading of such substances, such as chemically treated sand, can also remove an oil spill from sight and prevent its reaching shore. Dispersants are generally more effective on young spills, while sinking agents are more effective on the older, denser slicks.

Interior Department and Environmental Protection Agency regulations and those of other federal, state, and local agencies, plus the oil industry's recently increased awareness of the need for environmental protection, have caused a large market in spill-coping technology to appear. The ingenuity of American industry, of both the giant oil companies and small research firms, has been applied to developing better ways to prevent and control and cope with oil spills.[85] Therefore, the ability to contain and recover spills, to disperse or sink them with less harm to marine life, and to combat oil on water with microbial seeding and gelling and other new techniques, can be expected to improve over time.

NOTES

1. MIT Offshore Group, *Georges Bank Study,* 1973, Vol. II, p. 263.

2. For the interested reader a list of about 150 references bearing on effects of oil pollution is contained in: MIT Offshore Group, *Georges Bank Study,* 1974, Vol. II, pp. 252-262, 291.

3. Note Table 4-2, p. 49 above.

4. Frederick S. Hillier and Gerald L. Lieberman, *Introduction to Operations Research* (San Francisco: Holden-Day, Inc., 1967), pp. 38, 48, 79, 287, 628, 294.

5. MIT Offshore Group, *Georges Bank Study,* 1973, Vol. II, pp. 10-12, 17, 20.

6. For a more detailed discussion of petroleum components see: MIT Offshore Group, *Georges Bank Study,* 1973, Vol. II, pp. 184-202.

7. David P. Hoult, "Physical Effects of Oil on Aqueous Surfaces," in Robert W. Homles and Floyd A. DeWitt, editors, *Santa Barbara Oil Symposium* (Santa Barbara, Calif.: University of California at Santa Barbara, 1971).

8. William D. Garrett, "Impact of Petroleum Spills on the Chemical and Physical Properties of the Air-Sea Interface," in Holmes and DeWitt, eds., *Santa Barbara Oil Symposium,* 1971, pp. 89-90.

9. Craig L. Smith and William G. MacIntyre, "Initial Aging of Fuel Oil Films on Water," in American Petroleum Institute, Environmental Protection Agency, and U.S. Coast Guard, *Prevention and Control of Oil Spills* (Washington: American Petroleum Institute, 1971).

10. *Ibid.,* pp. 457-461.

11. This was the case in the West Falmouth and Tampico Maru spills (Max Blumer, "Scientific Aspects of the Oil Spill Problem," *Environmental Affairs,* I (April, 1971), pp. 54-73).

12. MIT Offshore Group, *Georges Bank Study,* 1973, Vol. II, p. 54.

13. Juris Vagnars and Paul Mar, *Oil On Puget Sound* (Seattle: University of Washington Press, 1972), p. 457.

14. Garrett, "Impact of Spills," in Holmes and DeWitt, eds., *Santa Barbara Oil Symposium,* 1971, p. 92.

15. Max Blumer et al., "The West Falmouth Oil Spill," (Unpublished manuscript, Woods Hole Oceanographic Institution Reference No. 70-44, Woods Hole, Mass., September 1970), p. 3.

16. Interview with Howard Sanders, Woods Hole Oceanographic Institution, Feb. 2, 1973.

17. M.H. Horn, J.M. Teal, and R.H. Backus, "Petroleum Lumps on the Surface of the Sea," *Science,* April 10, 1970, pp. 246-7.

18. This was the fate of some Torrey Canyon patches (J.E. Smith, ed., *"Torrey Canyon" Pollution and Marine Life* (Cambridge, U.K.: Cambridge University Press, 1968).

19. Large emulsion patches from the Torrey Canyon spills were menacing the French Coast more than a month after discharge of the oil (*Ibid.,* p. 157) and Horn reports dating some tar balls as being two months old (Horn, "Lumps on the Sea," *Science,* 1970, p. 246).

20. Max Blumer and J. Sass, "The West Falmouth Oil Spill, Data Available in November, 1971," (Unpublished manuscript, Woods Hole Oceanographic Institution Reference No. 72-19, Woods Hole, Mass., April 1972).

21. Garth D. Gumtz, *Restoration of Beaches Contaminated by Oil* (Washington: Office of Research and Monitoring, Environmental Protection Agency, September 1972), p. 7.

22. Much of this discussion is based on a number of conversations with scientists at Woods Hole Oceanographic Institution, especially with David Ross, and on his book: David A. Ross, *Introduction to Oceanography* (New York: Meredith Corporation, 1970).

23. A highly readable source on shore life is: Rachel Carson, *The Edge of the Sea* (New York: New American Library of World Literature, Inc., 1959).

24. Lyle S. StAmant, "Biological Effects of Petroleum Exploration

and Production in Coastal Louisiana," in Senate, *Outer Continental Shelf Policy Issues, Hearings,* 1972, pp. 1198-1249.

25. MIT Offshore Group, *Georges Bank Study,* 1973, Vol. II, p. 204.

26. For a fourteen-page-long table summarizing hundreds of toxicity tests on oil and marine species see: *Ibid.,* pp. 216-229.

27. Sources for this discussion of toxic effects are: *Ibid.,* pp. 203-236; R.B. Clark, "Oil Pollution and Its Biological Consequences," Report prepared for the Australian Great Barrier Reef Petroleum Drilling Royal Commissions (Great Britain: University of Newcastle upon Tyne, May 1971), pp. 17-28; and E.B. Cowell, ed., *The Ecological Effects of Oil Pollution on Littoral Communities* (London: Institute of Petroleum, 1971), p. 236.

28. A study of effects of oil on lobsters shows no effects observable when 10 ppm. of the water-soluble fraction of crude oil was added to water containing lobsters and a fifteen second delay in starting to search for food when 10 ppm. whole crude was added. These results indicate that if there is an effect it is small (J. Atema and Lauren Stein, "Sublethal Effects of Crude Oil on Lobster Behavior," (Unpublished manuscript, Woods Hole Oceanographic Institution Report 72-74, Woods Hole, Mass., September 1972). The MIT Offshore Oil Task Group (*Georges Bank Study,* 1973, Vol. II, p. 234) implicates levels of 10-100 parts per billion of oil in possibly disrupting such chemical communication, based on other studies, but this seems suspect as this is lower than the ambient level of hydrocarbons in coastal waters.

29. Max Blumer, G. Souza, and J. Sass, "Hydrocarbon Pollution of Edible Shellfish by an Oil Spill," (unpublished manuscript, Woods Hole Oceanographic Institution Reference No. 70-44, Woods Hole, Mass., January 1970). Max Blumer et. al., "West Falmouth Spill," (unpublished, 1970). Max Blumer and J. Sass, "West Falmouth Spill Data in 1971," (unpublished, 1972), and conversations with Howard Sanders, John Farrington, Frederick Grassle, and George Hampson at Woods Hole Oceanographic Institution, 1973.

30. Dale Straughan and R.L. Kolpack, editors, *Biological and Oceanographic Survey of the Santa Barbara Channel Oil Spill, 1969-1970.* (Los Angeles: Allan Hancock Foundation, University of Southern California, 1971).

Alfred W. Ebeling et. al., *Santa Barbara Oil Spill: Macroplankton, Fishes* (Washington: Interior Department, Federal Water Pollution Control Administration, 1969).

California Department of Fish and Game, California State Fisheries Laboratory, *Pelagic Fish Survey of Santa Barbara Oil Spill* (Cruise Report 69-A-4, Sacramento: Department of Fish and Game, 1970).

31. Smith, ed., *"Torrey Canyon" Pollution,* 1968.

A.J. Sullivan and A. J. Richardson, "The 'Torrey Canyon' Disaster and Intertidal Marine Life," *Nature* 214 (1967), pp. 541-542.

32. W.J. North, M. Neushul, and K.A. Clendenning, "Successive Biological Changes Observed in a Marine Cove Exposed to a Large Spillage of Mineral Oil," in *Symposium Pollu. mar. Micro-org. Prod. Petrol.* (Monaco, 1965), pp. 325-354.

33. G.L. Chan, *A Study of the Effects of the San Francisco Oil Spill*

on Marine Organisms (Kentfield, Calif.: College of Marin, 1971).

34. John G. Mackin, testimony in Senate, *Outer Continental Shelf Policy Issue, Hearings,* 1972, pp. 975-978.

35. *Ibid.,* p. 978.

36. St. Amant, "Biological Effects of Petroleum Production," in *Senate, Outer Continental Shelf Policy Issues, Hearings,* 1972, p. 1213.

37. Clark, "Oil Consequences," Great Britain, 1971, pp. 59-74.

38. These conclusions are also based on field trips to the sites of the West Falmouth spill, the Santa Barbara spill, and the Coal Oil Point natural oil seep north of Santa Barbara.

39. Max Blumer, "Scientific Aspects of the Oil Spill Problem," *Environmental Affairs,* I (April 1971), pp. 54-73.

40. *Ibid.,* p. 58.

41. MIT Offshore Group, *Georges Bank Study,* 1973, Vol. II, p. 235.

42. Interview with George Harvey, Woods Hole Oceanographic Institution, February 1973.

43. Claude E. ZoBell, "Sources and Biodegradation of Carcinogenic Hydrocarbons," in American Petroleum Institute, *Prevention of Oil Spills,* 1971, pp. 441-448.

44. M.J. Suess, "Polynuclear Aromatic Hydrocarbon Pollution in the Marine Environment," Food and Agriculture Organization Technical Conference on Marine Pollution and Its Effects on Living Resources and Fishing (Rome:FAO, 1970).

45. Clark, "Oil Pollution Consequence," (Great Britain, 1971), pp. 85-87.

46. The problem of when to take substances off the market because they contain some level of pesticides or oil or other chemical that is definitely harmful at high concentrations has become complex because improved analytical techniques can detect concentrations as low as a few parts per billion. At such levels it is nearly impossible to separate natural hydrocarbons from those added by man's activities.

47. Horn, "Petroleum Lumps," *Science,* April 10, 1970, pp. 246-7.

48. Blumer, "Scientific Aspects of the Oil Spill Problem," *Environmental Affairs* (April 1971), p. 58.

49. Smith, ed., *"Torrey Canyon" Pollution,* 1968, pp. 151-162.

Henry G. Schwartzberg, "The Movement of Oil Spills," in American Petroleum Institute, *Prevention of Oil Spills,* 1971.

Hoult, "Effects of Oil on Aqueous Surfaces," in Holmes and De-Witt, eds., *Santa Barbara Oil Symposium,* 1971.

MIT Offshore Group, *Georges Bank Study,* 1973, Vol. II, pp. 60-1.

50. Information for this discussion come primarily from interviews with Dean Bumpus, Woods Hole Oceanographic Institution, 1973, and from: Dean F. Bumpus and Louis M. Lauzier, *Surface Circulation on the Continental Shelf Off Eastern North America between Newfoundland and Florida,* Foilo 7 of the *Serial Atlas of the Marine Environment* New York: American Geographical Society, 1965).

MIT Offshore Group, Georges Bank Study, 1973, Vol. II, pp. 62-68.

Dean F. Bumpus, "A Description of the Circulation on the Continental Shelf of the East Coast of the United States," (Unpublished manuscript, Woods Hole Oceanographic Institution, Woods Hole, Mass., 1972).

51. MIT Offshore Group, *Georges Bank Study*, 1973, Vol. II, pp. 60-100. (Those without computers can get less quantitative results using geometry on a map of the area.)

52. From the 200 simulations in spring 1/200 or .005 reached shore, in summer 10/200 or .05 reached shore.

53. Burton T. Coffey, "Fishing Agreements Prove of Small Value," *National Fisherman*, May 1973, p. 2-A.

54. Sources for this discussion have been numerous interviews with scientists of the National Marine Fisheries Service and a large number of articles during the past three years in national newspapers and in the *National Fisherman*.

55. Elements of this discussion are taken from: Henry B. Bigelow and William W. Welsh, *Fishes of the Gulf of Maine*, Bulletin of the United States Bureau of Fisheries, Vol. XL (Washington: Government Printing Office, 1925). Henry B. Bigelow, *Plankton of the Offshore Waters of the Gulf of Maine*, Bulletin of the Bureau of Fisheries, Vol. XL, Part II (Washington: Government Printing Office, 1925). Henry B. Bigelow, Lois C. Lillick, and Mary Sears, *Phytoplankton and Planktonic Protozoa of the Offshore Waters Gulf of Maine* (Philadelphia: American Philosophical Society, 1940).

56. MIT Offshore Group, *Georges Bank Study*, Vol. 1, pp. 195-7.

57. *Ibid.*, pp. 174-6.

58. *Ibid.*, pp. 211-17.

59. *Ibid.*, pp. 222-33.

60. *Ibid.*, pp. 231-3.

61. As with catches of other species on the Bank, sea scallop catches are in precipitous decline, with the 1971 catch one-third that of 1961. Canada and the U.S. are the only harvesters. (MIT Offshore Group, *Georges Bank Study*, 1973, Vol. I, p. 1950.)

62. This seems reasonable, taken from a map of scallop catch areas (*Ibid.*, p. 197).

63. The net value of the damage to fishermen would be much less since not all the scallops could be caught and since boat and labor costs are saved if the area is not fished.

64. See above, section 4D.

65. P. Korringa, "Biological Consequences of Marine Pollution with Special Reference to North Sea Fisheries," *Helgolander wiss. Meeresunters*, 17 (1968), pp. 126-40.

H.A. Cole, "North Sea Pollution," United Nations Food and Agriculture Organization Conference on Marine Pollution and Its Effects on Living Resources and Fishing Paper No. R-20 (Rome: FAO, 1970).

66. Clark, "Oil Pollution Consequences," Great Britain, 1971, p. 79.

67. *Ibid.*

68. For a detailed analysis of this aspect of offshore operations on Georges Bank see: MIT Offshore Group, *Georges Bank Study*, 1973, Vol. I, pp. 198-210.

69. *Ibid.*, p. 198-9.

70. Interior Department, Final Environmental Statement, 1972, p. 55.

71. National Petroleum Council, *Environmental Conservation,* 1972, Vol. II, p. 212.

72. The data on which this discussion is based are from: MIT Offshore Group, *Georges Bank Study,* 1973, Vol. I, pp. 80-94.

73. The West Falmouth Spill was from a #2 fuel barge load.

74. For extensive and detailed analysis of pipeline, tanker and barge, offshore platform, and storage oil spills see: MIT Offshore Group, *Georges Bank Study,* 1973, Vol. II, pp. 1-50, and Dillingham Corporation, *System Study of Oil Cleanup Procedures* (La Jolla, Calif.: Dillingham Corp., March 1970).

75. This is three-fourths of all imports; the remaining quarter comes overland from Canada. Data on imports and production are from: British Petroleum Company, *Statistical Review,* 1971, pp. 18-22.

76. Interior Department, *Outer Continental Shelf Statistics,* 1972, pp. 75-76.

77. MIT Offshore Group, *Georges Bank Study,* Vol. I, pp. 37-8.

78. Imported oil is expected to fill the gap between supply and demand at current prices and even if gas prices are deregulated and climb (see above, Chapter 3, and MIT Offshore Group, *Georges Bank Study,* Vol. I, pp. 136-9).

79. From Table 2-6, 10 billion cu. ft. gas equals .24 million tons of oil.

80. For analysis of impacts of the Santa Barbara spill on recreators and the tourist industry see: Walter J. Mead and Philip E. Sorensen, "The Economic Cost of the Santa Barbara Oil Spill," in Holmes and DeWitt, eds., *Santa Barbara Oil Symposium,* 1971.

81. This is not a detailed or extensive analysis of spill coping methods both because they are not significantly applicable to spills far offshore and because the analysis has been done elsewhere. A few references are:

Dillingham Environmental Company, *Systems Study of Oil Spill Cleanup Procedures,* 1970.

Battelle Memorial Institute, *Oil Spill Treating Agents – A Compendium,* May 1970.

American Petroleum Institute and Federal Water Quality Administration, *Proceedings of Joint Conference on Prevention and Control of Oil Spills* (New York: American Petroleum Institute, 1970).

American Petroleum Institute, Environmental Protection Agency, and U.S. Coast Guard, *Prevention and Control of Oil Spills* (Washington, American Petroleum Institute, 1971).

National Petroleum Council, *Environmental Conservation,* Vol. II, pp. 257-267.

82. *Ibid.,* pp. 257-8.

83. *Ibid.,* p. 259.

84. Detailed discussion of this possibility is contained in: MIT Offshore Group, *Georges Bank Study,* 1973, Vol. II, pp. 120-60.

85. Fruits of these labors are evident in the trade journals *Ocean Industry, Oceanology, Undersea Technology,* the *Oil and Gas Journal,* and others.

Chapter Five

Major Findings and Recommendations Based on Economic and Environmental Considerations

SUMMARY OF FINDINGS

The estimates of possible outcomes associated with either Georges Bank petroleum production or the importation of the equivalent amount of oil from the Middle East are summarized in Table 5-1. The three numbers under many of the outcome categories refer to outcomes resulting from production of the low, medium, and high Georges Bank petroleum production estimates.

There probably are commercial petroleum deposits under the outer half of Georges Bank. On the basis of finds in similar sediments off Nova Scotia, it can be surmised that any petroleum beneath Georges Bank most likely is in the form of gas and the light fractions of oil. Oil production would probably be less than 60,000 barrels a day but might range up to 400,000. Potential gas production over forty years could be about 400 million cubic feet a day and might range up to 2,000 million. The medium estimate of potential daily oil and gas production is the British Thermal Unit equivalent of about 100,000 barrels of oil, the high estimate of about 560,000. If Georges Bank petroleum is not produced, this equivalent amount of oil would be imported from the Middle East. Georges Bank petroleum might displace about one percent of U.S. oil imports in the 1980s. If extremely large deposits are found, it might displace up to five percent.

Georges Bank petroleum would most likely be moderate cost oil and gas. That is, it would be produced at costs well above production costs in the Middle East but well below the expected U.S. price of oil and well below the cost of shale oil, imports of liquified natural gas, and of alternate sources of energy. Landed costs of Georges Bank oil might be $1.50-$2.00 a barrel, about $2.50-$3.00 after taxes and royalties. Since the late 1972 price of oil landed at the East Coast averaged about $3.75 and all predictions are for steady increases, Georges Bank production would probably be profitable. Only if international oil prices plummet by a dollar or more and unrestricted oil imports are permitted

109

Table 5-1.   Summary of Possible Outcomes

Sets of three numbers refer to possibility of low-medium-high Goerges Bank oil and gas production estimates.

Georges Bank Production   OR

oil   0 – 60,000 – 400,000 bbl/day
gas   0 –   400 –   2,000 million cu ft/
                                          day

Displaces % of U.S. Oil Imports in mid-1980s
0 – 1% – 5%

U.S. Government Receives:

   lease payments
   0 – $700 million – $4 billion
   annual royalties
   0 – $22 million – $126 million

U.S. Production Inputs Receive Annually
–$60 million+

Annual Expenditures in New England
0 – $15 million – $75 million

Possible New Jobs in New England
0 – 140 – 600

Major Oil Spills on Georges Bank 40 Year Period

size	number
–10,000 bbl	0 – 6 – 70
1,0000-100,000	0 – 2 – 25
100,000 bbl	0 – 1 – 10

Probability of One or More Spill Remnants Reaching Southeast New England Shore – 40 Year Period

size	probability
–10,000 bbl	0 – .05 – .50
10,000–100,000	0 – .05 – .25
100,000 bbl	0 – .05 – .10

Damage to Fishery:

From oil spills-negligible.

From chronic discharge of oil–"worst case"–up to 2% of sustainable sea scallop harvest contaminated.

Oil Imports from Middle East
0 – 100,000 – 560,000 bbl/day

Middle East Government Taxes and Royalties/Year
0 – $80 million – $450 million

Foreign Tanker Operator Receives Payments/Year
0 – $20 million – $112 million

U.S. Government Duty/Year
0 – $ 8 million – $42 million

Short Run Net Outflow of Dollars/Year (in millions)

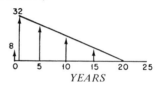

Georges Bank – continued

Transport of Georges Bank Petroleum Production

Change in Risk of Nearshore Oil Spills to New England in mid-1980s

Oil Shipped to New England
0 – 0 – 0

Oil Piped to New England Refinery
0 – -1% – -4%

Oil Piped to Shore and Shipped Out of Region
0 – +3% – +18%

Gas Piped to New England
0 – -1% – -3%

into the United States—an unlikely combination of events— would Georges Bank production most likely not be worth producing.

Should the decision be made to lease Georges Bank the federal government could receive $700 million from the oil industry in lease payments; they could range as high as $4 billion. Once production starts, royalty payments of $22 million could be paid to the federal government if production is the equivalent of about 100,000 barrels of oil a day. If high estimates for both oil and gas production are brought to the surface, royalties could be $126 million a year. The Georges Bank lessees would spend about $60 million a year in the United States to produce the oil and gas and to transport it to shore. If the equivalent amount of oil is imported from the Middle East, the foreign government would receive $80 million in taxes and royalties and the foreign tanker owners and operators would receive more than $20 million a year. The federal government would receive about $8 million annually in duty payments, $42 million if 560,000 barrels are imported.[1] Therefore, if Georges Bank petroleum is produced, the immediate dollar payments accrue to the American factors of production, the federal government, and the producing and transport companies. If imports are needed to compensate for foregone Goerges Bank production, the immediate dollar payments go primarily to the Middle East governments and the foreign oil tanker operators.

Such dollar movements, in the case of imports, would result in a short run net outflow which would affect the U.S. balance of payments. The first year investment in the Middle East to create the capacity to export 100,000 barrels of oil a day, plus the daily purchases of this amount, might cause a net annual outflow of about $8 million the year of investment, about $32 million the first full year of importing the 100,000 barrels, and of decreasing amounts for the next twenty or so years. By that time the lagged return flows of dollars to the U.S. by way of third countries would roughly equal the annual outflow, and the net effect on the balance of payments would be negligible. The short-run net outflows, however, would come at the end of a very large deficit in the U.S. balance of payments in fuels. The increased dollar outflow would add to an already large inflow of dollars and other currencies from industrialized nations to Middle Eastern governments, the nations who own most of the world's petroleum reserves.

If Georges Bank development is permitted, annual expenditures in New England for production and transportation operations could be $15 million and might be as high as $75 million. Such operations might add 140 jobs or, in the event production is extremely large, 600 new jobs. If there is no oil and gas in commercially producible amounts under Georges Bank—a possibility— there would be few economic impacts on New England.

If oil is found and produced on Georges Bank, it would introduce the new hazard of major oil spills originating there. Over the entire possible forty year oil production period the probability is that there would be no large spill or one large spill of more than 100,000 barrels of oil, but there might be as many as ten or more. There would also probably be two or fewer spills of 10,000-100,000 barrels, with the possibility of up to twenty-five or more. And there would most likely be six or fewer small spills of less than 10,000 barrels, although the number could range up to seventy or more.

The shore areas most probably affected by one of these spills would be southern and western Cape Cod, Nantucket, Martha's Vineyard, and Rhode Island beaches. But during the forty year period the probability of one or more spills in each size category reaching these shores is less than five percent. With extremely large-scale production of oil on Georges Bank the risk of one or more aged spills reaching shore might be ten percent for large ones, twenty-five percent for medium-sized ones, and about fifty percent for small ones over a forty year period. A spill would not impact shore until it had been at sea at least six weeks. By that time most components of the oil toxic to marine life would have evaporated or dissolved, and the remnants of the spill would most likely be in the form of widely scattered small tar balls and patches of oil in water emulsion. Such remnants would have a small temporary adverse effect on shore marine life. But they could antagonize people who either see or hear about or step on them.

Major oil spills could remain on the Bank for a number of weeks. However, the Georges Bank marine ecosystem is so prolific and resilient and protected from local shocks that infrequent large oil spills or localized chronic discharges of low concentrations of oil would have, at worst, minimum adverse effects on marine species in the area. The long-run effects of oil spills seem negligible because the oil is naturally dispersed and diluted by enormous quantities of water and is eventually oxidized and biologically degraded into simpler, generally harmless compounds. Recovery and repopulation of oil-contaminated areas does occur, although the process may take years in heavily contaminated shallow waters and in bottom sediments.

The major stress to the Georges Bank fishery is the overfishing by large foreign factory ship and trawler fleets. In a few years stocks of commercially valuable species may be depleted below the possibility of recovery. This seems to have already occurred with haddock. International management of the fishery has failed. Hope for sustainable long-run productivity rests with the possibility of single nation management, which is a plausible outcome of the 1974 United Nations Law of the Sea Conference. Should this occur, the worst credible damage to the fishery could result from chronic low level discharges of oil in wastewater if ten large oil production platforms are placed on the sea scallop grounds on the southwestern part of the Bank. In this "worst case," with the high estimate of production all being produced there, up to two percent of the annual sustainable sea scallop harvest might be contaminated with oil and rendered commercially unusuable.

The transportation to shore of Georges Bank oil and gas could increase or decrease or leave unchanged the total risk of nearshore oil spills to the New England Coast. Nearly all New England oil consumption is carried to ports there in tankers and barges from refineries aboard or on other U.S. coastal areas. If total oil tonnage moved along the cost is considered a proxy for the risk of nearshore oil spills, the shipping of Georges Bank oil by tanker to New England would result in no change in the risk of spills, because it would simply displace the equivalent amount of tonnage carried from other areas. If Georges Bank oil is carried by pipeline to shore, the risk of nearshore oil spills could be decreased by about one percent and perhaps by as much as four percent because pipeline transport of oil is about forty percent safer than sea shipment. That is, forty percent fewer spills can be expected from pipeline than from surface transport of the same amount of oil. However, if Georges Bank oil is piped to New England and then transshipped out of the region for refining, the total tonnage moved along the coast would increase, and the risk of spills might be increased by three percent and possibly as much as eighteen percent. Bringing Georges Bank gas to New England would displace imported oil, thus reducing the risk of nearshore oil spills by one percent to three percent. Therefore, a net increase in the risk of nearshore oil spills to the New England Coast from transport of Georges Bank oil and gas would occur only if large amounts of oil are brought to New England and then shipped out of the region. In the other cases it appears that the total risk would decrease. This concludes the summary of major findings.

Recommendations

If more federal revenue is preferred to less, and if payments to United States workers and investors are preferred to payments to Middle East governments and foreign tanker operators, then with respect to these criteria it is desirable to lease Georges Bank for oil and gas development from a national point of view. The effect of such a decision on the risk of nearshore oil spills to the New England Coast could range from an increase of up to eighteen percent to a decrease of up to seven percent. If the oil from Georges Bank does not have to be shipped out of New England to a refinery, the impact on the region's environment would be favorable since the Georges Bank oil and gas would displace an equivalent amount of coastal oil tanker and barge traffic to New England ports. The environmental risks associated with offshore production operations would most likely be negligible. Therefore, from the New England region's point of view, the new environmental impacts associated with Georges Bank petroleum development would be small. Since the development would possibly add a few hundred new jobs and annual expenditures of millions of dollars to the region's economy, the decision to lease would also appear to be desirable from the New England regional standpoint.

Petroleum development on Georges Bank could come in conflict with use of the area as a fishery. However, the adverse impacts on fishery productivity would be negligible, a conclusion also reached by the MIT Offshore Oil Task Group.[2] Only about a fifth of the Georges Bank catch is made by United States fishermen. Most of the remainder is being taken by large foreign fishing fleets that seem to care little for the long-run productivity of the fishery. Therefore, the potential small effects on the fishery do not appear reason enough to forego petroleum development on Georges Bank.

When to hold lease sales also must be considered. Each year the lease sales are delayed is a year in which oil payments are made to Middle Eastern governments and foreign tanker operators. It is also a year of foregone federal revenue. If a dollar in hand is preferable sooner than later, if preventing dollar outflows from the United States is preferable sooner than later, and if creating jobs and new investment opportunities in the United States is preferable sooner than later, then Georges Bank should be leased as soon as possible.

There are a number of factors that act as constraints on the timing of OCS lease sales, making it advisable to stagger sales to avoid low bids. Among these are limits to industry capital available for lease bidding, availability of exploratory drilling rigs and well development equipment, and availability of trained workers. There are also legal and political considerations which are discussed in the following chapter.[3]

NOTES

1. The size of this would increase if the tariff is raised from the 1975 rate of $0.21 per barrel and appropriates any further increases in prices of imported oil.

2. "We couldn't identify any environmental effect associated with a well-run offshore oil production operation." (J.W. Devanney III, Oral Presentation on *The Georges Bank Petroleum Study,* Massachusetts Institute of Technology, Cambridge, Mass., March 22, 1973).

3. Several policy matters relating to OCS leasing and development have not been discussed in this report because they apply to all OCS areas. They bear on Interior Department administration of the Outer Continental Shelf Lands Act of 1953 rather than on the Georges Bank decision in particular. Among these issues are: (1) separating leases into exploratory drilling leases and development leases so the government assumes the risk of no shows rather than the oil companies, (2) giving small independent operators more of an opportunity to participate in OCS leasing, (3) regulating allowable production, (4) preserving oil reserves for future medical and chemical uses, (5) making company acquired geologic information about OCS areas public instead of proprietary.

Chapter Six

Legal and Political Considerations

LEGAL CONSIDERATIONS

The Boundary Disputes

Boundary disputes make up a very large chapter in the history of Interior Department leasing of the OCS. Most of the boundary between Louisiana and federal submerged lands remained unresolved for fifteen years as cases dragged on in the Supreme Court.[1] Boundary disputes have developed between the states of Alaska and California and the federal government. In 1969 the State of Maine sold petroleum exploration leases to King Resources, a private development firm, for an area more than 100 miles from the Maine coast.[2] The Justice Department, on information from Interior, brought suit against Maine to halt such leasing beyond the three mile limit.[3] The Submerged Lands Act of 1953 granted coastal states the title to submerged lands out to their traditional boundaries, three miles from their shore for most states. Texas received a three-league boundary, 10.5 miles, because that distance was part of its constitution when the Republic of Texas entered the Union. Florida received three leagues on its Gulf Coast because that boundary was recognized by Congress when Florida re-entered the Union after the Civil War. All other states have three mile limits;[4] California, Alabama, Mississippi, and Louisiana, have lost their attempts before the Supreme Court to get more. The federal government, then, has jurisdiction over the seabed beyond the state boundaries to the outer limit of national jurisdiction, presently the 200 meter depth contour.[5]

Massachusetts has claimed a seabed boundary out to 200 miles to try to protect its fishing industry from foreign trawlers. The "Home of the Cod" state and other states claim that their colonial charters granted them jurisdiction over the ocean floor 200 miles from the coastline, which includes Georges Bank. Eleven other Atlantic Coast states have joined Maine and Massachusetts. In 1970 the Supreme Court appointed a special master, The Chief Judge of the Third Circuit Court of Appeals in Philadelphia, to hear testimony and make recommendations prior to the Court's considering the case. He has said he will make his report no earlier than 1974.[6] If either party disagrees with his recommendations the boundary issue could be before the Supreme Court for a number of years.

While the suit is pending, the Interior Department cannot lease tracts in the disputed area unless an agreement is made with the states involved— in the case of Georges Bank, with Massachusetts. Seismic exploration in the area has occurred during the court dispute, and Interior granted permits for the activity. The Massachusetts Attorney General has warned the exploration companies that they have been proceeding in areas under dispute.[7] An arrangement could be made to hold any leasing and other revenues in escrow and to hold joint-lease sales. Something like this arrangement was made between Louisiana and Interior in 1956, permitting sales and development to progress into the 1970s while cases were being considered by the Supreme Court.

In the absence of a court decision or of an agreement between the states and Interior, a lease sale of tracts on Georges Bank cannot legally be held.[8] It is beyond the scope of this report to analyze the merits of the case on both sides. However, the facts that the state claims did not come until sixteen years after passage of the Submerged Lands Act, and that past Supreme Court decisions have found in favor of the historic three mile limit as state seafloor boundaries, indicate that it is likely the states will lose their extended claims. If Massachusetts does win a 200 mile boundary, however, the decision to lease would be up to the state, and any revenues would accrue to the state instead of to the federal government.

Another party, Canada, also disputes federal claims to Georges Bank. The Ottawa government claims about one-third of Georges Bank, the eastern third, under the 1958 Law of the Sea treaty which provided a formula for delineating international boundaries. Since 1965 the Canadian Department of External Affairs has granted geological exploration permits (no drilling) for this part of the Bank. Washington challenged the validity of these permits in 1969, and the two countries held two rounds of inconclusive talks in 1970.[9] The international boundary issue will probably be resolved at the international Law of the Sea Conference to be held in Santiago, Chile in 1974. If the boundary rule adopted there does give Canada one-third of the Bank, any revenues from petroleum development would, of course, accrue to Canada.

NEPA

The second legal consideration relates to the National Environmental Policy Act of 1969. This act, referred to as NEPA, requires all federal agencies to publish an environmental impact statement for every major action that might affect the environment. The detailed statement, to be prepared by the responsible official, must include all environmental impacts of the proposed action, adverse environmental effects which cannot be avoided, and alternatives to the action. Guidelines for preparation of such statements have been promulgated by the Council on Environmental Quality (CEQ).

The first general lease sale in the Gulf of Mexico that was part of President Nixon's 1971 Clean Energy Message program to accelerate OCS leasing was held up for nearly a year by a lawsuit involving NEPA. The suit, brought by the Natural Resources Defense Council and other environmental litigating groups, successfully claimed that all the alternatives to the action, including alternatives not under Interior's jurisdiction, were not considered in the impact statement.[10] The U.S. District Court for the District of Columbia held that the broad policy of accelerating OCS leasing should have been accompanied by an environmental impact statement. Since this had not been done by President Nixon and his Domestic Council, the agency first to implement a part of the policy was required to formulate the comprehensive statement. In this case, the job fell to Interior. The first sale under the acceleration policy, scheduled for December, 1971, was cancelled. Interior then added sections to the statement,[11] released it, and the sale was held in September of 1972. This environmental impact statement and one for a subsequent sale in December, 1972, appear to be legally adequate. They were not challenged in court.

The Bureau of Land Management, the responsible agency in Interior which originates the impact statements, and other agencies and offices in Interior are now able to produce legally acceptable environmental impact statements for OCS lease sales. Many personnel are involved and many manhours, but the NEPA requirements are now an administrative burden on Interior rather than a constraint that might prevent a lease sale.[12]

POLITICAL CONSIDERATIONS

The Executive Branch

Support for accelerated leasing of the OCS for oil and gas development, including off the Atlantic Coast, has been nearly unanimous among high officials of the Nixon Administration, including the President himself. He called for such accelerated leasing in his 1971 Clean Energy Message,[13] and he directed the Interior Department, in his April, 1973, Energy Message, to take steps to triple the annual acreage leased on the OCS by 1979.[14] The statutory decision maker for leasing OCS areas is the Secretary of the Interior, currently Rogers C.B. Morton. But a decision to go ahead with Atlantic OCS leasing, once the boundary dispute is settled in favor of the federal government, would most likely be a collegial decision. The cast of officials involved in energy policy formulation has been changing every other month. The director of the Energy Policy Office in the White House and the Secretary of the Department of Energy and Natural Resources, if that department is created, would play key roles. It is not necessary to assess how the decision might be made in the Nixon Administration to lease Georges Bank because nearly all the top officials would favor it. They generally want increased development of secure domestic sources of energy, reductions in the balance of payments deficits, and increased revenues to the Treasury from lease sales and royalty payments.[15] Few, if any, of them have strong ties to the Northeast where opposition to Georges Bank leasing is heavy. Therefore, they can be expected to support the leasing of Georges Bank as soon as possible. Of course if the lease sales have not occurred by 1977, a new President, especially if he is from New England, might have different thoughts on petroleum development off the East Coast.

The three federal agencies that might raise objections to leasing Georges Bank are the Environmental Protection Agency, the Council on Environmental Quality, and the National Oceanic and Atmospheric Administration, which contains the National Marine Fisheries Service. The last is charged with protecting U.S. fishery interests. These agencies are not in the decision-making chain between the President and the Secretary of the Interior; their only formal channel for influencing the decision is through commenting on the environmental impact statement. This might result in small changes in implementing the decisions to lease and develop Georges Bank for oil and gas production—that is, a few tracts might be withheld because of potential damage to scallops or other marine resources—but these agencies would be unlikely to attempt to prevent a decision to lease Georges Bank.

Congress

The leasing of Georges Bank would be an administrative decision made by the Secretary of the Interior. It would require no new authorization, and holding a lease sale would not require any new appropriations. The subsequent supervision of the lessees would require additional appropriations of a few hundred thousand dollars for the U.S. Geological Survey. The authorization and appropriations committees in Congress with jurisdiction over OCS policies are made up primarily of Western development-oriented Congressmen who favor domestic sources of oil and gas over foreign ones and favor increases in federal revenue from OCS leasing. Consequently, they can be expected to support the leasing of Georges Bank. By and large, congressmen from East Coast states, who might oppose the leasing, are not on committees that could directly influence the decision.

Support for Leasing

Other support for leasing Georges Bank comes from the oil industry and from development-oriented state and local organizations in New England.[16] Energy-using utilities and industries also generally support oil and gas production there. The shortages of heating fuel during the 1972-73 winter and the gasoline shortages in the summer of 1973, plus the numerous advertisements and television commercials sponsored by oil companies and the American Petroleum Institute, may be creating an attitude favorable to Georges Bank leasing among segments of the general public. This support complements the attitudes of the Republican federal officials involved in making the decision.

Opposition to Leasing

Opposition to possible leasing has already arisen. Environmentalists, fishermen, and some public officials have already expressed hostility to any petroleum development off the New England and New York shores.[17] East Coast state governors have met with Secretary Morton to express their concern about possible environmental damage.[18] A comparatively favorable position has been taken by Massachusetts Governor Francis W. Sargent, who told the Senate Interior and Insular Affairs Committee that he is "not completely opposed" to the idea of drilling oil wells off the New England Coast.[19] Although opponents have little direct influence over a Nixon administration decision to lease—unless, of course, the Supreme Court decides in favor of the states—state and local officials could deny pipeline rights-of-way across state owned submerged lands and could make acquisition of shore support and terminal facilities difficult. Such actions might not be able to prevent development, but they could make operations more expensive for the lessees.

There are a number of reasons for this opposition to Georges Bank petroleum development. The first is the not unsurprising fact that those to benefit directly from the petroleum development and those to lose are generally different parties. Beneficiaries would be the federal government, the domestic oil industry, and suppliers of services and equipment and capital to offshore oil and gas production and transportation operations. The possible losers are a wide class of people since the damage could affect parts of a long section of coastline. Therefore, those who have shore property, who live near the shore, who use the shore, or who simply care about the shore and the sea, feel the risk of damage from oil spills should be avoided. Fishermen would bear the risks of damaged marine resources and fouled gear while obtaining no benefits. State officials see the federal government receiving the lion's share of the revenues from the petroleum development while their constituents bear the new risks. It seems natural that the "losers" would oppose the leasing.

Another factor that facilitates opposition is that the decision to lease Georges Bank is seen by itself as a yes or no decision instead of as the alternative to an increment in tanker imports from the Middle East. In such a framework the disadvantages of Georges Bank development stand all by themselves instead of being compared with the disadvantages of increased tanker traffic to the United States and increased tanker and barge oil traffic in and out of New England ports.

A third factor is the continued strength of concern for the environment as a public issue. Downs estimates that concern for the environment is about midway through the "issue-attention cycle" and that, as an issue, this concern may remain at a comparatively high level for a long time.[20] The issue may have already gone beyond the stage of "alarmed discovery and euphoric enthusiasm" and into the stages of "realizing the costs of significant progress" and "gradual decline of intense public interest." Downs conjectures that the environmental issue may not soon decline to the "post-problem stage," where there is little more than spasmodic interest in the subject, for a number of reasons. First, many forms of environmental pollution, especially air and water pollution, are clearly visible and repugnant to many and can threaten nearly everyone. Second is that attacking environmental pollution is generally not politically divisive, i.e., it is not attacking the stakes of large blocks of voters. Third, a large share of the blame for environmental degradation, especially the more visible kind, can be placed on small groups of powerful "villains," turning efforts to make them clean-up into popular crusades against wealthy despoilers. Fourth, through pressure from government officials, those responsible for pollution can actually be made to prevent it and clean it up. Therefore, successes are possible. Fifth, the cost of successes can be spread out in prices and taxes, means of passing on costs that are not highly visible.[21] All these factors apply to oil pollution from offshore operations. The possible perpetrators of environmental damage are large oil companies, who are already suitable targets for criticism due to high fuel costs in New England, to the oil import quotas which have prevented less expensive oil from reaching New England, and to the depletion allowance and other tax breaks granted the industry.

An important factor contributing to opposition is that many people believe oil spills cause a great amount of damage. This is because an oil spill, particularly one on shore, is a newsworthy and photogenic event. It can cause serious short-term damage, killing birds and fish and blackening beaches and boats. This provides the drama for a news story, and conflict can be added by interviewing an outraged shellfish harvester and a spokesman for the company responsible. The story, with its drama and conflict, can be tied into a national theme running through society—in the case of oil spills, to the increasing price paid by Americans for a damaged environment, or to other similar themes.[22] The events communicated to the public by reporters are those with the most attention-getting value. This is to be expected because, as Downs states, "A problem must be dramatic and exciting to maintain public interest because news is 'consumed' by much of the American public largely as a form of entertainment."[23] Therefore, the immediate damage of an oil spill is presented, often sensationalized, while damage that did not occur and the gradual recovery of the area receive must less conspicuous attention. Many news reporters have been at the forefront in concern for protecting the environment and have made strong efforts to highlight serious pollution episodes. The media attention understandably gives the impression to the public that oil spills are more damaging than is demonstrated here in Chapter Four. Furthermore, speculation about catastropic effects from pollution, most of which have very low probabilities of being true, such as Blumer's cancer hazard and predictions of the oceans dying, receive coverage while findings of complicated or small or no damaging effects receive no coverage or, frequently, coverage that is distorted to make the findings newsworthy.

Finally, many people have personal attitudes that cause them to see Georges Bank petroleum development as undesirable in a larger scheme of things. Some of these attitudes are: distrust of government, antipathy to technology, antipathy to the oil industry, a desire to decrease use of energy, a desire to live in tune with nature, and dislike of the Nixon Administration. For all these reasons, public opposition to any lease sale of Georges Bank tracts can be expected to be vocal and widespread.

Dealing with Opposition

Several possibilities are available to the Interior Department to allay fears and reduce opposition to leasing and petroleum development on Georges Bank. Some of these are:

> 1. To require that the lessees have available at all times enough dispersants and sinking agents to insure that no oil spill from Georges Bank could reach shore.
> 2. To provide opportunities to New England officials, fishermen, and environmentalists to participate in overseeing lessee compliance with pollution prevention regulations.
> 3. To inform the media on the findings contained in this book, and to hold discussion periods with reporters. In this case, it appears that the more analysis and information available the better.

In the case of a policy decision which may benefit a few parties to a large extent and harm others to a small extent, some form of compensation is a useful mechanism for obtaining widespread support for the decision. Such compensation would automatically occur, at least for outdoor recreators, from this decision because royalties and other revenues from OCS oil and gas production enter the Land and Water Conservation Fund, administered by the Interior Department, which finances the acquisition of parkland by federal, state, and local agencies. In Fiscal Year 1972, OCS revenues made up $285 million of total Fund receipts of $361 million.[24] Consequently, recreators and environmentalists are compensated, nationwide, for damage sustained by marine areas from OCS petroleum operations.

A more direct mode of compensation would be actual sharing of OCS revenues with the adjacent coastal states, a proposal enthusiastically endorsed by Louisiana, Texas, and California officials. This direct sharing of revenues from federally owned lands with adjacent states has been recommended by the Public Land Law Review Commission[25] and is now advocated by some New England spokesmen.[26] Such a policy would clearly benefit the adjacent coastal states, making federal leasing more attractive to their officials, while

85849

depriving the federal Treasury and perhaps the Land and Water Conservation Fund of a like amount of revenue for nationwide benefits.

How much of the opposition might be mitigated by various federal efforts is uncertain. A large part of it is and will remain implacable, born of many precedents and deep frustrations and intense concern for an unspoiled environment. But things can be done. Public hearings on the lease sale might be conducted as part of the process of formulating an environmental impact statement. Further efforts to inform the opponents, to take some of their advice, and above all to involve them in the regulation of lessees—all these might go a short distance toward making the climate a more friendly one for all involved.

Notes

1. Senate, *Governmental Intervention, Petroleum Industry, Hearings,* 1970, Part 5, pp. 1908-2016.

2. James Ayres, "Offshore Drilling Battled in Legislature, Courts," Boston Globe, June 25, 1972, p. 2.

3. U.S. v. Maine, et. al., Supreme Court, Original, 1969.

4. These state titles are to the sea floor. The water above is under federal jurisdiction.

5. There is confusion about this boundary. The 1958 Geneva Convention on the Continental Shelf gave seabed jurisdiction to the coastal nation out to the 200 meter depth contour or to the limit of ability to exploit natural resources. This ambiguity may be cleared up in the forthcoming International Law of the Sea Conference commencing in November 1973.

6. David A. Andelman, "East Coast Wary of Offshore Oil," *New York Times,* April 12, 1973, p. 53.

7. Robert E. Cahill, "On the Horns of a Dilemma," *Yankee,* January 1973, pp. 94-148.

8. U.S. Department of the Interior, Office of the Secretary, "Secretary Morton Says No Decision Made or Now Planned to Lease Atlantic Coast Outer Continental Shelf Oil and Gas," Interior Department News Release, November 4, 1972.

9. Ken O. Botwright, "US, Mass. Head for Ocean Oil Collision," *Boston Sunday Globe,* April 29, 1973, p. 58.

10. Source for this discussion is: "NRDC v. Morton: Significant New Appellate Decision on Section 102 of NEPA," *102 Monitor,* Council on Environmental Quality, February 1972, pp. 1-20.

11. Interior Department, *Final Environmental Statement Proposed 1972 Lease Sale,* 1972.

12. Senate, *National Environmental Policy Act, Joint Hearings,* 1972, testimony by Secretary Morton and other Interior officials.

13. U.S., President, "The President's Message to the Congress, June 4, 1971," *Weekly Compilation of Presidential Documents,* Vol. 7, No. 23, June 7, 1971, pp. 855-66.

14. "Energy Message Excerpts," *New York Times*, April 19, 1973, p. 53.

15. In mid-1973 such officials include Treasury Secretary Schultz and Energy Policy Office Director John Love.

16. "Con Ed Chief Urges Offshore Drilling," *New York Times*, October 10, 1972, p. 71.

Edward Cowan, "Oil Industry Opens Drive for Sea Drilling," *New York Times*, July 23, 1972.

Cahill, "Dilemma," *Yankee*, January 1973.

R.S. Kindleberger, "Panel Calls for Gas, Oil Exploration off N.E.," *Boston Globe*, December 8, 1972, p. 23.

Gene Smith, "Utilities to Search Ocean for Natural Gas Supplies," *New York Times*, March 26, 1973, p. 63.

17. Ed Forry, "Offshore Oil Exploration Plans Meet Legislative Resistance," *Jamaica Plain Citizen*, July 6, 1972, p. 2.

William M. Bulger, "Another View on Offshore Oil Drilling," *Boston Globe*, December 11, 1972, letter to editor.

Gladwin Hill, "Santa Barbara, 3 Years Later, Fights Oil Drilling," *New York Times*, January 29, 1972, p. 27.

David A. Andelman, "Drilling Off L.I. Foreseen as Peril," *New York Times*, March 13, 1972, p. 26.

18. January 11, 1972 at the Interior Department.

19. "Offshore Oil Use Mulled by Sargent," *Boston Globe*, February 2, 1973, p. 32.

20. Anthony Downs, "Up and Down with Ecology—the 'Issue-Attention Cycle,' " *The Public Interest*, Summer 1972, pp. 38-50.

21. *Ibid.*, pp. 46-48.

22. This coverage is standard for TV network news (Edward J. Epstein, "The Selection of Reality," *The New Yorker*, March 3, 1973, pp. 41-77).

23. Downs, "Up and Down with Ecology," *Public Interest*, p. 42.

24. "Congress Creates but Doesn't Appropriate," *Audubon*, March 1973, p. 114.

25. Public Land Law Review Commission, *One Third of the Nation's Land* (Washington: Government Printing Office, June 1970).

26. R. Frank Gregg, Chairman of the New England River Basins Commission, panel discussion on Georges Bank development, Massachusetts Institute of Technology, Cambridge, Mass., March 22, 1973.

Sources Consulted[1]

GOVERNMENT PUBLICATIONS

Bigelow, Henry B. and William W. Welsh. *Fishes of the Gulf of Maine.* Bulletin of the United States Bureau of Fisheries, Vol. XL, Part I. Washington, D.C.: Government Printing Office, 1925.

Bigelow, Henry B. *Plankton of the Offshore Waters of the Gulf of Maine.* Bulletin of the Bureau of Fisheries. Vol. XL. Part II. Washington D.C.: Government Printing Office, 1925.

California Department of Fish and Game. California State Fisheries Laboratory. *Pelagic Fish Survey of Santa Barbara Oil Spill.* Sacramento, Calif.: Cruise Report 69-A-4, Department of Fish and Game, 1969.

Ebeling, Alfred W. et. al. *Santa Barbara Oil Spill: Macroplankton, Fishes.* Washington, D.C.: Interior Department, Federal Water Pollution Control Administration, November 1969.

Gumtz, Garth D. *Restoration of Beaches Contaminated by Oil.* Report EPA-R2-72-045. Washington, D.C.: Environmental Protection Agency, 1972.

International Commission on Northwest Atlantic Fisheries. *Statistical Bulletin for 1970.* Vol. 20. Dartmouth, Nova Scotia, Canada: International Commission on Northwest Atlantic Fisheries, 1972.

Moore, J. Cordell. *United States Petroleum Through 1980.* Washington, D.C.: Department of the Interior, 1968.

Public Land Law Review Commission. *One Third of the Nation's Land.* Washington, D.C.: Government Printing Office, 1970.

Uchupi, Elazar. *Atlantic Continental Shelf and Slope of the United States-Physiography.* Geological Survey Professional Paper 529-C. Washington, D.C.: Government Printing Office, 1968.

[1]Newspaper articles, short magazine articles, and minor sources footnoted in the text are not reproduced here. The source for many impressions and observations has been regular reading of *The New York Times, The Boston Globe, Time, Ocean Industry, Oil and Gas Journal, National Fisherman,* and other periodicals.

Uchupi, Elazar. *Atlantic Continental Shelf and Slope of the United States-Shallow Structure.* Geological Survey Professional Paper 529-I. Washington, D.C.: Government Printing Office, 1970.

U.S. Cabinet Task Force on Oil Import Control. *The Oil Import Question.* Washington, D.C.: Government Printing Office, 1970.

U.S. Department of Commerce. *Climatological and Oceanographic Atlas for Mariners. Vol. I. North Atlantic Ocean.* Washington, D.C.: Department of Commerce, 1959.

U.S. Department of the Interior. Geological Survey. *Outer Continental Shelf Statistics.* Washington, D.C.: Geological Survey, 1972.

U.S. Department of the Interior. *Petroleum and Sulfur on the U.S. Continental Shelf.* Washington, D.C.: Department of the Interior, 1969.

U.S. Department of the Interior. Bureau of Land Management. *Final Environmental Statement Proposed 1972 Outer Continental Shelf Oil and Gas General Lease Sale Offshore Eastern Louisiana.* Washington, D.C.: Department of the Interior, June 20, 1972.

U.S. Department of the Interior. Bureau of Land Management. *Draft Environmental Statement Proposed 1972 Outer Continental Shelf Oil and Gas General Lease Sale Offshore Louisiana.* Washington, D.C.: Department of the Interior, July 1972.

U.S. Department of the Interior. Geological Survey. *Outer Continental Shelf Lease Management Study: Safety and Pollution Control.* Washington, D.C.: Geological Survey, 1972.

U.S. Department of the Interior. Bureau of Land Management. *The Role of Petroleum and Natural Gas from the Outer Continental Shelf in the National Supply of Petroleum and Natural Gas.* Washington, D.C.: Government Printing Office, May 1970.

U.S. Federal Power Commission. Bureau of Natural Gas. *National Gas Supply and Demand 1971-1990.* Washington, D.C.: Government Printing Office, 1972.

CONGRESSIONAL HEARINGS

U.S. Congress. House. Committee on Interior and Insular Affairs. *Policies, Programs, and Activities of the Department of the Interior. Hearings* before the Committee on Interior and Insular Affairs, House of Representatives, 91st Cong., 1st sess., 1969.

U.S. Congress. Senate. Committee on the Judiciary. *Governmental Intervention in the Market Place. The Petroleum Industry. Hearings* before a subcommittee of the Committee on the Judiciary, Senate, 91st Cong., 2nd sess., 1970.

U.S. Congress. House. Committee on Interior and Insular Affairs. *Santa Barbara Channel Leases, California. Hearings* before a subcommittee of the Committee on Interior and Insular Affairs, House of Representatives, on H.R. 18159 and related bills, 91st Cong., 2nd sess., 1970.

U.S. Congress. House. Committee on Appropriations. *Department of the Interior and Related Agencies Appropriations for 1973. Hearings* before a subcommittee of the Committee on Appropriations, House of Representatives, 92nd Cong., 2nd sess., 1972.

U.S. Congress. Senate. Committee on Public Works and Committee on Interior and Insular Affairs. *National Environmental Policy Act. Joint Hearings* before the Committee on Public Works and the Committee on Interior and Insular Affairs, Senate, 92nd Cong., 2nd sess., 1972.

U.S. Congress. Senate. Committee on Interior and Insular Affairs. *Outer Continental Shelf Policy Issues. Hearings* before the Committee on Interior and Insular Affairs, Senate, 92nd Cong., 2nd sess., 1972.

BOOKS, PROCEEDINGS, REPORTS

Adelman, M.A., *The World Petroleum Market.* Baltimore: John Hopkins University Press, 1972.

American Petroleum Institute and Federal Water Pollution Control Administration. *Proceedings of the Joint Conference on Prevention and Control of Oil Spills.* Dec. 15-17, 1969. New York: American Petroleum Institute, 1970.

American Petroleum Institute, Environmental Protection Agency, and U.S. Coast Guard. *Prevention and Control of Oil Spills.* Washington, D.C.: American Petroleum Institute, 1971.

Battelle Memorial Institute. *Oil Spill Treating Agents–A Compendium.* Battelle Memorial Institute, 1970.

Bigelow, Henry B.; Lillick, Lois C.; and Mary Sears. *Phytoplankton and Planktonic Protozoa of the Offshore Waters Gulf of Maine.* Philadelphia: The American Philosophical Society, 1940.

British Petroleum Company Limited. *BP Statistical Review of the World Oil Industry – 1971.* London: British Petroleum Company, 1972.

Bumpus, Dean F. and Louis M. Lauzier. *Surface Circulation of the Continental Shelf Off Eastern North America between Newfoundland and Florida.* New York: American Geographical Society, 1965.

Carson, Rachel. *The Edge of the Sea.* New York: New American Library of World Literature, Inc., 1959.

Carthy, J.D. and Don A. Arthur, eds. *The Biological Effects of Oil Pollution on Littoral Communities.* Pembroke, Wales, England: Field Studies Council, 1968.

Chan, G.L. *A Study of the Effects of the San Francisco Oil Spill on Marine Organisms.* Part I. Kentfield, Calif.: College of Marin, 1971.

Cowell, E.B., ed. *The Ecological Effects of Oil Pollution on Littoral Communities.* London: Institute of Petroleum, 1971.

Dillingham Environmental Corporation. *Systems Study of Oil Cleanup Procedures.* La Jolla, Calif.: Dillingham Corporation, 1970.

Holmes, Robert W. and Floyd A. DeWitt, Jr., eds. *Santa Barbara Oil Symposium.* Dec. 16-18, 1970. Santa Barbara, Calif.: University of California at Santa Barbara, 1971.

Massachusetts Institute of Technology. Offshore Oil Task Group.

The Georges Bank Petroleum Study. 2 Volumes. Report No. MITSG 73-5. Cambridge, Mass.: Massachusetts Institute of Technology Sea Grant Program, 1973.

National Petroleum Council. *Environmental Conservation the Oil and Gas Industries.* 2 Volumes. Washington, D.C.: National Petroleum Council, June 1971 and February 1972.

Ross, David A. *Introduction to Oceanography.* New York: Meredith Corporation, 1970.

Smith, J.E. ed. *"Torrey Canyon" Pollution and Marine Life.* Cambridge, United Kingdom: Cambridge University Press, 1968.

Straughan, Dale and R.L. Kolpack, eds. *Biological and Oceanographic Survey of the Santa Barbara Channel Oil Spill, 1969-1970.* Los Angeles: Allan Hancock Foundation, University of Southern California, 1971.

Vagnars, Juris and Paul Mar. *Oil on Puget Sound.* Seattle, Washington: University of Washington Press, 1972.

MAGAZINE AND JOURNAL ARTICLES, PAPERS

Adelman, M. "Is the Oil Shortage Real? Oil Companies as OPEC Tax Collectors." *Foreign Policy,* 10 (December 1972).

Blumer, Max. "Scientific Aspects of the Oil Spill Problem." *Environmental Affairs,* I (April 1971), 53-74.

Bonafede, Don. "White House Report/President Nixon's Executive Reorganization Plans Prompt Praise and Criticism." *National Journal,* 5 (March 10, 1973), 329-344.

Business Week. "The Middle East Squeeze on the Oil Giants." July 29, 1972, pp. 56-62.

Cahill, Robert E. "On the Horns of a Dilemma." *Yankee,* January 1973, pp. 94-148.

Clark, R.B. "Oil Pollution and Its Biological Consequences." Report prepared for the Australian Great Barrier Reef Petroleum Drilling Commissions. Great Britain: University of Newcastle upon Tyne, May 1971.

Corrigan, Richard. "Energy Report/Demand for More Oil and Gas Prompts Review of Offshore Leasing." *National Journal,* 4 (July 8, 1972), 1109-16.

Corrigan, Richard. "Energy Report/Administration Readies 1973 Program to Encourage More Oil, Gas Production." *National Journal,* 4 (Oct. 21, 1972), 1621-32.

Downs, Anthony. "Up and Down with Ecology–the 'Issue Attention Cycle.' " *The Public Interest.* Summer 1972, pp. 38-50.

Emery, K.O. "Geology of the Continental Margin off Eastern United States," in *Proceedings of the Seventh Symposium of the Colston Research Society,* April 1965. London: Butterworths Scientific Publications, 1965.

Emery, K.O. and Elazar Uchupi. "Structure of Georges Bank." *Marine Geology,* 3 (1965), 349-58.

Epstein, Edward J. "The Selection of Reality," *The New Yorker.* March 3, 1973, pp. 41-77.

Faltemayer, Edmund. "The Energy 'Joyride' is Over." *Fortune,* September 1972, pp. 99-192.

Gardner, Frank J. "Vast Worldwide Trade Blooming in LNG." *Oil and Gas Journal.* Sept. 11, 1972, pp. 52-55.

Horn, M.H.; Teal, J.M.; and R.H. Backus. "Petroleum Lumps on the Surface of the Sea." *Science,* 168 (April 10, 1970), 246-7.

Kennedy, John L. "North Sea Plans Turned into Tangibles." *Oil and Gas Journal.* Jan. 8, 1973, pp. 65-9. "Oil and Gas Technology Offshore of the United Kingdom," George Williams, General Manager, Shell UK Exploration and Production Ltd., at Financial Times Seminar on the North Sea.

North, W.J.; Neushul, M.; and K.A. Clendenning. "Successive Biological Changes Observed in a Marine Cove Exposed to a Large Spillage of Mineral Oil." *Symp. Pollut. mar. Micro-org. Prod. Petrol.,* Monaco: 1965, pp. 325-354.

O'Sullivan, A.J. and A. J. Richardson. "The 'Torrey Canyon' Disaster and Intertidal Marine Life." *Nature,* 214 (1967), 541-2.

St. Amant, Lyle S. "Biological Effects of Petroleum Exploration and Production in Coastal Louisiana." Louisiana Wildlife and Fisheries Commission, Baton Rouge, Louisiana, December 1970.

Straguhan, Dale. "Factors Causing Environmental Changes After an Oil Spill." *Journal of Petroleum Technology,* 24 (1972), 250-4.

REGULATIONS

U.S. Code of Federal Regulations, Title 30. "Mineral Resources," Revised as of Jan. 1, 1972.

U.S. Code of Federal Regulations, Title 43. "Public Lands," Revised as of Jan. 1, 1972.

U.S. Department of the Interior. Geological Survey. Conservation Division. Branch of Oil and Gas Operations. Gulf Coast Region. *Notice to Lessees and Operators of Federal Oil, Gas, and Sulfur Leases in the Outer Continental Shelf Gulf Coast Region, OCS Orders Nos. 1 through 11.* New Orleans, Louisiana: Geological Survey, 1972.

U.S. Department of the Interior. Geological Survey. Conservation Division. Branch of Oil and Gas Operations. Pacific Region. *Notice to Lessees and Operators of Federal Oil and Gas Leases in the Outer Continental Shelf Pacific Region.* Los Angeles, Calif.: Department of the Interior, 1972.

INTERVIEWS

During the ten months this report was in preparation, June 1972–April 1973, interviews were conducted with officials in the Interior Department, particularly in the Geological Survey, and with scientists in the National Marine Fisheries Service. Interviews and discussions were also conducted with scientists at Woods Hole Oceanographic Institution.

Index

abalones: Tampico Maru, 72
Africa: West and North, 21
Adelman, M.: cost in Persian Gulf, 25, 31
Alaska: and California disputes, 115; oil, 17
Algeria: outflow returns, 29
American Petroleum Institute, 118
Amoco spill, 151
Army Corps of Engineers, 94
aromatics, 67, 76; definition, 57, 58

Bahamas: return flow, 29
balance of payments, 28, 30, 111; Georges
    Bank and imports, 31; return flows, 28
Baltimore Canyon, 2
Bay of Fundy, 83
beach sand, 61
benthic fauna, 65–69
Bermuda: return flow, 29
biodegradation: West Falmouth, 69
birds: and spills, 73
bivalves, 67
Blumen, M., 75, 76
bonus: payments, 34
boundary: disputes, 115
British Petroleum, 28
Brittany, 71
Bureau of Land Management, 117

Cabinet Task Force on Oil Import Control,
    28
Canada: Dept. of External Affairs, 116;
    gas pipeline, 21; refineries, 40; as
    source of oil, 21
Caribbean, 21
CEQ (Council on Environmental Quality),
    117, 118

Chase Manhattan Bank: 1971, 1972 con-
    sumption, 17
chemical communication, 104; and spills, 68
Clark, R.B., 73, 93
containment: devices, 101, 102
costs: definition, 36

decision-makers: probability and outcomes, 4
degradation, 60
Delaware Bay, 40
Dept. of Interior: leases, 2, 19; OCS orders,
    93; pipelines, 99
Dept. of Justice, 115
detergents, 71
Digicon, 8; exploration, 36
dollar flows, 23–26
Downs, A., 119, 120

economics: of off-shore oil, 5
eco-system: health of, 4
Energy Policy Office, 117
environment: birds and spills, 73; and off-
    shore oil, 5; spills, 55
Environmental Protection Agency, 118

fishing industry: international, 2
Florida: boundaries, 115
food chain, 75
FPC (Federal Power Commission), 19,
    38–41

gas, 19; chromatography, 69
gastropods, 68
Geneva Convention on the Continental
    Shelf, 1958, 122
Geological Survey, 9

Georges Bank: description, 2–7; fish, 86–89; impact, 41; oil cost, 25; oil slick path, 77; opposition, 121; potential production, 13; spill path prediction, 82
Global Marine, 36
Golden Gate Bridge crash, 72
Great South Channel, 79
Gulf of Maine, 78
Gulf of Mexico: OCS, 13, 25; oysters, 66
Gulf Oil, 28

Hendricks, T.A., 10
holes: exploratory, 8
Horn, M.H., 60
Hosmer, Craig, 1
Hoult, D.P., 58

ICNAF (International Commission for the Northwest Atlantic Fisheries), 86
import: cost breakdown, 27; and leases, 5; U.S. in 1972, 17
Indonesia, 23
International Law of the Sea Treaty, 1958, 116
International Law of the Sea Conference, 1973, 112, 122
International Law of the Sea Conference, 1974, 116
intertidal life: and Santa Barbara, 70; and spills, 86

Jersey Standard, 28

King Resources, 115

Land and Water Conservation Fund, 121
leases, 114; increase by 1979, 19; and opposition, 121; and revenue, 2
Libya, 23, 24
litigation: rights, 2
LNG (liquified natural gas), 21
Louisiana: offshore operation, 72; OLS, 23, 24; oil regulatory bodies, 42

McKelvey Estimates, 9
Mackin, J.G., 72
Mandatory Oil Import Program, 21
marine life, 58; feeding habits and metabolic rate, 65; and natural factors, 63; and water movement, 78
market removal, 105
Martha's Vineyard, 79, 98
Massachusetts: boundary, 115
media: and spills, 120
Middle East, 17–23; and Georges Bank, 109; prices, 33
M.I.T. Offshore Study Group, 75; chemical

communication, 104; and competitive bidding, 34; conclusions, 114; and development costs, 36; Georges Bank costs, 25; impact, 23, 91; laboratory experiments, 67; oil transportation costs, 38; pollution, 46; spill predictions, 82–85
Mobil Oil Co., 28
model: analysis, 5; of impact on environment, 46
Morton, Rogers C.B., 117

Nantucket, 98
Narragansett Bay, 79
National Environmental Policy Act, 1969, 116
National Oceanic and Atmospheric Administration, 118
National Petroleum Council, 101
nekton, 65–68
Nelson, T.W. and Burke, C.A., 10
Netherlands Antilles: return flow, 29
New England: consumption in 1985, 40; impact of Georges Bank, 35
NGL (natural gas liquids), 10
Nixon, R.M., 19; Clean Energy Message, 117
North Sea, 39
nuclear energy, 34

OCS (Outer Continental Shelf), 1
Odeco, 36
Oil Import Program: and U.S. prices, 25–33
Oil Policy Committee, 19–21
OPEC (Organization of Petroleum Exporting Countries), 26–30
Oregon: OCS, 11
Outer Continental Shelf Land Acts, 1953, 1, 114
overfishing, 67, 112
oxidation, 60

particles, 59
pelagic birds: and Santa Barbara, 70
phytoplankton, 65, 75
pipeline, 38
plankton, 65; Santa Barbara, 70; and Torrey Canyon, 71
platform, 36
policy analysis, 3
pollutants: other than oil, 66
production: and Georges Bank employment, 39; impact of Georges Bank, 22; manpower needs, 37; prediction of, 19; rate of, 12
Providence: refinery in, 40
Public Land Law Review Commission, 121

quota: oil import, 21

refineries, 40
residue, 60
return flows, 29, 30
Royal Dutch Shell, 28

Santa Barbara Channel, 76; leases, 13; spill, 51, 70
Sargent, F.W., 119
Saudi Arabia, 23, 24; cost breakdown, 27; outflow returns, 29; price of a barrel, 25
sea lions: Santa Barbara, 70
seals, 70
sea urchins: Tampico Maru, 72
sediment, 9; and pollution, 59; Santa Barbara, 70
seepage: Santa Barbara, 71
shipyards, 28
Smith, C.L. and MacIntyre, W.G., 58
spills, 47, 55; aged, 62; and beaches, 112; and birds, 73; and components, 55; coping, 101, 102; damage to marine life, 66; net impact, 4; phases of aftermath, 69; risks, 99; size and probability, 52, 53; spreading, 58
St. Amant, L.S., 73
Standard of California, 28
stripper wells, 25
Spivak, J. and Shelburne, O.B., 11
Submerged Lands Act, 115, 116
subtidal organisms, 65
Suess, M.J., 75
Supreme Court, 119; rights to resources, 2

Switzerland: return flow, 29

Tampico Maru, 72
tankers, 94-98
tariff, 21, 114
Task Force: return flow, 30
taxes: OPEC, 33
Tehran Agreement, 33
Texaco, 28
Texas: boundaries, 115; oil regulatory bodies, 42
The Offshore Company, 36
Torrey Canyon, 71
transportation, 95-97; mode and pollution, 48; projection, 100, 101

U.S. Bureau of Mines, 17
U.S. State Department, 34
U.S. Treasury, 35

Venezuela, 21-24; outflow returns, 29

Washington OCS, 11
wastes, 93
water movement: and pollution, 78
West Falmouth, 61, 69
whales, 70
wind rose, 80

ZoBell, C.E., 75
zooplankton, 65

About the Author

William R. Ahern, Jr. received his Ph.D. in Public Policy from Harvard University where his major fields were environmental impact assessment and natural resource management. He is presently a policy analyst with the Energy Program of the Rand Corporation. Dr. Ahern has spent a year at Woods Hole Oceanographic Institution as a Fellow in Marine Policy and Ocean Management and another year as a management intern with the Department of the Interior. His B.S. is from the Air Force Academy.